Mechanical Engineering Series

Frederick F. Ling
Series Editor

Springer
New York
Berlin
Heidelberg
Barcelona
Budapest
Hong Kong
London
Milan
Paris
Singapore
Tokyo

Mechanical Engineering Series

(continued after index)

Anthony Lawrence

Modern Inertial Technology

Navigation, Guidance, and Control

Second Edition

With 201 Figures

 Springer

Anthony Lawrence
76 Glen Street
Whitman, MA 02382, USA

Series Editor
Frederick F. Ling
Ernest F. Gloyna Regents Chair in Engineering
Department of Mechanical Engineering
The University of Texas at Austin
Austin, TX 78712-1063, USA
    and
William Howard Hart Professor Emeritus
Department of Mechanical Engineering,
    Aeronautical Engineering and Mechanics
Rensselaer Polytechnic Institute
Troy, NY 12180-3590, USA

Library of Congress Cataloging-in-Publication Data
Lawrence, Anthony, 1935–
    Modern inertial technology : navigation, guidance, and control /
Anthony Lawrence — 2nd ed.
        p.   cm. — (Mechanical engineering series)
    Includes bibliographical references and index.
    ISBN 0-387-98507-7 (hardcover : alk. paper)
    1. Inertial navigation (Aeronautics)   I. Title.   II. Series:
Mechanical engineering series (Berlin, Germany)
TL588.5.L38   1998
629.132′51—dc21   98-13047

Printed on acid-free paper.

Production managed by Anthony K. Guardiola; manufacturing supervised by Jeffrey Taub.
Camera-ready copy prepared from the author's WordPerfect files.
Printed and bound by Edwards Brothers, Inc., Ann Arbor, MI.
Printed in the United States of America.

9 8 7 6 5 4 3 2 1

ISBN 0-387-98507-7 Springer-Verlag New York Berlin Heidelberg   SPIN 10658368

Series Preface

Mechanical Engineering, an engineering discipline borne of the needs of the industrial revolution, is once again asked to do its substantial share in the call for industrial renewal. The general call is urgent as we face profound issues of productivity and competitiveness that require engineering solutions, among others. The Mechanical Engineering Series features graduate texts and research monographs intended to address the need for information in contemporary areas of mechanical engineering.

The series is conceived as a comprehensive one that covers a broad range of concentrations important to mechanical engineering graduate education and research. We are fortunate to have a distinguished roster of consulting editors on the advisory board, each an expert in one of the areas of concentration. The names of the consulting editors are listed on the next page of this volume. The areas of concentration are applied mechanics, biomechanics, computational mechanics, dynamic systems and control, energetics, mechanics of materials, processing, thermal science, and tribology.

I am pleased to present this volume in the Series: *Modern Inertial Technology: Navigation, Guidance, and Control*, Second Edition, by Anthony Lawrence. The selection of this volume underscores again the interest of the Mechanical Engineering series to provide our readers with topical monographs as well as graduate texts in a wide variety of fields.

Austin, Texas   Frederick F. Ling

Mechanical Engineering Series

Frederick F. Ling
Series Editor

Preface

Since 1993, when the first edition of this book was published, inertial technology has changed in two ways. First, the maturing of the Global Positioning System (GPS) has encouraged electronics manufacturers to produce simple, inexpensive ($100) position indicators for the general public. Also, silicon micromachined gyroscopes and accelerometers have come of age and are now mass-produced. Together, these developments have impacted the low-cost, low-accuracy inertial system market.

Secondly, the Interferometric Fiber Optic Gyroscope (IFOG) has become a reliable and accurate sensor and has found a market in heading and attitude reference systems. Different IFOG technologies have converged to a fairly standard instrument.

In this second edition, we have generally updated each chapter and expanded the text and references relating to the micromachined sensors and the IFOG. While we cannot describe some proprietary design features, there is enough public literature available so that the reader can understand recent technological advances.

We decided not to remove descriptions of some of the older technology (floated gyros, for example), as these may well be in the inventory for years to come. Also, the Pendulous Integrating Gyroscope Accelerometer (PIGA), based on this technology, has not yet been bettered as a precise accelerometer, although engineers are still attempting to make a "modern," solid-state, less expensive, and more reliable replacement.

There were a few errors in the first edition that have been corrected. Our thanks to those who took the time to point them out.

Whitman, MA   Anthony Lawrence

Contents

Introduction

Automatic navigation makes ocean-going and flying safer and less expensive: Safer because machines are tireless and always vigilant; inexpensive because it does not use human navigators who are, unavoidably, highly trained and thus expensive people. Unmanned deep space travel would be impossible without automatic navigation. Navigation can be automated with the radio systems Loran, Omega, and the Global Positioning System (GPS) of earth satellites. In some circumstances, such as when a submarine is deeply submerged or in a war zone where radio signals may be jammed, these aids are not available. Then we must use a self-contained system called *inertial navigation.*

Inertial navigation uses gyroscopes and accelerometers (inertial sensors) to measure the state of motion of the vehicle by noting changes in that state caused by accelerations. By knowing the vehicle's starting position and noting the changes in its direction and speed, one can keep track of the vehicle's current position. Mankind first used this technology in World War II, in strategic guided weapons where cost was unimportant; only 20 to 30 years later did it become cheap enough to be used commercially.

The electronics revolution, in which vacuum tubes were replaced by integrated circuits, dramatically altered the field of inertial navigation in two ways: the deployment of GPS has swung the balance in favor of aided systems, and microelectronics has altered the technology of inertial systems. Early inertial systems used complex mechanical gimbal structures and mechanical gyroscopes with spinning wheels. The gimbals allowed the gyroscopes to stabilize a body (called a *platform*) so that it remained in a fixed attitude relative to a chosen coordinate frame, even as the vehicle turned around any or all of its three major axes. Nowadays, the gimbals are often replaced by a computation scheme that uses the gyros' measurements of the vehicle angles to calculate where the vehicle axes are with respect to the chosen coordinate frame. These computations cannot be done in real time without a fast digital computer, an important product of the electronics revolution. Also, modern gyros are likely to use light beams (perhaps from a laser) rather than wheels, circulating light instead of circulating metal.

Without the electronics revolution, space satellites would not be possible, and the relatively low cost of launching satellites has given us the GPS radio aid mentioned earlier. These satellites are navigation beacons that provide electronic signals for special radio receivers to decode, thereby locating the navigator anywhere on the earth's surface. GPS is important because it gives much higher location accuracy than Loran and Omega, to tens of meters. If the user is stationary, perhaps when surveying, specialized GPS receivers can provide position to centimeter accuracy.

The combination of GPS signals and inertial navigation provides inexpensive, precise navigation for the GPS bounds the errors of the inertial navigator and the latter keeps navigating if the satellite signals should be temporarily obscured (by hills or tunnels, or jammers in war).

Navigators need to know the time accurately. As part of its position fixing signals, GPS provides the accurate time. Also, quartz crystal clocks developed in the last 20 years are so accurate and inexpensive that anyone can know the time, to fractions of a second, for only a few dollars.

As a spin-off, quartz crystals similar to timekeeping oscillators can make inexpensive, good accelerometers. And the technology of integrated circuits allows micromachined silicon devices to be made, up to a hundred to a wafer, providing alternate types of miniature accelerometers and gyros. The modern technology of guided wave optics, fiber optics (developed for telecommunications), and integrated optics enables cheaper, more reliable gyros to be made.

This avalanche of technologies begun by the electronics revolution has resulted in lower-cost inertial navigation systems. Fiber-optic gyros can be found in aircraft such as the Boeing 777. The new technology is extending the application of inertial navigation to munitions. Bombs are getting smarter, and cannon shells can be guided to their targets. Lower accuracy GPS receivers are already available at a reasonable price—a few hundred dollars—for the weekend yachtsman, and automobile manufacturers offer navigation systems in their more luxurious models.

The purpose of this book is to present the technology of the gyroscopes and accelerometers used in inertial navigation so that the reader can understand how they work, the advantages and disadvantages of the different types available, and where the field is headed.

It should be useful to systems engineers who specify and select navigation, guidance, and control systems and sensors; project managers seeking background knowledge to help them to sift the wheat from the chaff when listening to competing claims (not always objectively presented) from equipment manufacturers; new engineers of all ages beginning or moving into a career in the field; engineers and scientists deeply immersed in one area of the field who wish to comprehend the scope of the whole; and the curious reader, schooled in science, who wants to know how inertial sensors work.

While most of the work described in the following chapters was done in the United States (because that is work that the author knows best), much good inertial system and component work has been done in Europe (and in what was the USSR, if their space program is any guide). With the advent of optical gyros, though, we are seeing much innovation from Japan. I have tried to reflect this swing in the references selected for each chapter.

Very few technical papers were written on mechanical gyros, probably due to security restrictions in the 1960–'70s. Ring laser gyros were more thoroughly covered, but still the total volume of publication in that field is modest. Now, the fiber-optic gyros have been so extensively written about that it is almost impossible to keep up—570 papers were listed in a 1989 bibliography!

There is someone who should be singled out for his success in promulgating the international dissemination of the lore of inertial technology. That man is Dr. Helmut Sorg, of the University of Stuttgart, Germany. Every year, for about 20 years, he organized the "Symposium Gyro Technology" for the *Deutsche Gesellschaft für Ortung und Navigation* (DGON; roughly "German Organization for Guidance and Navigation"), collecting papers (in English) into proceedings that are invaluable for the study of the field. I have referred to them liberally, and I recommend them to you.

1
An Outline of Inertial Navigation

Navigation's Beginnings

Pioneers returning from their journeys provided travel instructions for those who wished to repeat their journeys. They wrote descriptions of their routes and made charts or maps pointing out landmarks and hazards such as rivers and mountains on land or shoals and rocks at sea. Mapmakers devised a global coordinate system using a grid of latitude and longitude circles, by which the position of any place on earth could be defined.

Buoys and lighthouses provided route markers for sailors close to land, but once they ventured into the featureless seas their pathfinding became much more difficult. Magnetic compasses were known as early as the eleventh century, but were not very accurate. The fourteenth-century navigator used the stars to find latitude, and by the eighteenth century, technology had provided instruments like the sextant for more accurate measurement of the positions of the celestial bodies.

Finding longitude was more difficult, because the position of the celestial bodies depends on the rotational position of the earth relative to the stars, that is, the time of day. Longitude is measured from an arbitrarily chosen zero at Greenwich, near London, England; to determine longitude, navigators need to know the time where they are and the time at Greenwich (Greenwich Mean Time, GMT). They can find local time from observations of the sun, but finding GMT means observing the satellites of Jupiter, or lunar distances [1], which are uniquely positioned at any time. But these observations require complicated calculations to yield time, so it is much better to have a clock set to GMT on board. It took the invention of the spring escapement clock in the mid-eighteenth century to make accurate timekeeping available at sea. How the eighteenth-century mariner would envy today's $10 quartz watch, accurate beyond his wildest dreams! Nowadays, GMT is known as UTC, or Coordinated Universal Time; it is broadcast by radio, world-wide, and is available in the United States over telephone lines from the National Institute of Standards and Technology and the Naval Observatory.

This process of following a path on a map through predetermined latitudes and longitudes is called *navigation*, and the process of pointing the vehicle to follow a chosen path is called *guidance*. Navigation at sea uses the process of *deduced reckoning* (called "dead" reckoning), illustrated in Figure 1.1. If a ship starts from

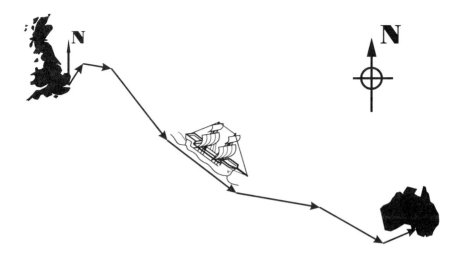

Figure 1.1. Inertial navigation is dead reckoning.

a known latitude and longitude and travels in a known direction for a known time, its position on the previously charted sea is known and can be verified by sun (or other celestial body) sightings. As long as the ship's direction and speed are known accurately, the sailors will land in the correct harbor.

When aircraft began to fly over the sea, they got direction and speed from the magnetic compass and airspeed indicator. As radar evolved, the Doppler set could give more accurate ground speed. Dead reckoned courses tried to follow planned routes, corrected by celestial navigation, but airplanes fly so fast that the navigator was hard pressed to make celestial measurements quickly enough. More accurate and safer navigation was possible when land-based radio beacons and systems like Loran and Omega became available.

By 1908, the magnetic compass had been replaced by the gyrocompass. Gyrocompasses are important marine navigation instruments, but they are not generally useful for aircraft, spacecraft, or guided missiles. Because our focus is on the instruments for inertial navigators, we are not going to consider the gyrocompass in this book; it is described in general navigation books such as Dutton's [2], by Wrigley in the *Encyclopedia Britannica* under "Gyrocompass," and by others [3, 4].

Military aircraft prefer not to depend on radio beacons in time of war because an enemy can jam them, and aircraft can be detected by the enemy if they transmit Doppler radar speed measurement signals. They need a self-contained navigation system. If one wants to guide unmanned vehicles (such as a space probe or an intercontinental missile), one must also have a self-contained navigation system. Civil aircraft need to have reliable navigation at the lowest possible cost, and human navigators are expensive. The solution to all these needs is to automate navigation, which means that one must find a way to keep track of position, speed, and attitude continuously, in all weather conditions, in space, in the sky, and under the sea. Inertial navigation serves this purpose.

Inertial Navigation

Gyroscopes and accelerometers can provide the necessary signals for automatic navigation. Gyroscopes measure rotation, and accelerometers measure acceleration. Integrating the output from an accelerometer gives speed, and integrating speed gives distance traveled. The gyroscopes provide information on where the accelerations are directed, and therefore heading and distance, the essential ingredients for dead reckoning, are known. As these instruments use the inertial properties of matter (or of light) for their operation, dead reckoning with gyros and accelerometers is called *inertial navigation* (IN).

The first inertial navigators were used in the German V1 and V2 weapons in World War II. After the war, a group of German scientists, under Werner Von Braun, developed this technology at Redstone Arsenal in Huntsville, Alabama, for ICBMs and spacecraft, building inertial navigation systems for U.S. Army missiles.

Other U.S. groups developed IN systems, notably one under Draper at the East-coast Massachusetts Institute of Technology [5, 6]. Their first aircraft inertial navigators were flown in 1949, followed in 1954 by the Navy Ship's Inertial Navigation System. On the West coast, Autonetics made inertial systems using a different design philosophy. In 1958, the nuclear-powered submarine *Nautilus* successfully navigated under the polar ice past the North Pole using an Autonetics XN6 navigator. In the 1960s the Apollo program took inertial guidance into space, and now inertial systems are widely used, even being used in "smart" munitions. For more on the history of inertial navigation, see Wrigley's summary [7] and Mackenzie's book [8].

Maps and Reference Frames

Before we go anywhere, we need to orient ourselves in the space in which we will navigate; we do this with *maps* and *reference frames*. A map or chart is drawn to some scale so that the user can calculate distances between places. Over the distance covered by a town map we assume that the flat sheet represents a flat area of the earth, and although there may be hills and valleys in the town, the town is represented as if it were a model on a flat board. But for larger areas, because the earth is roughly a sphere, it cannot be precisely represented by a flat map. Many different projections have been developed for making flat navigation maps [2], but we will let those readers who are interested pursue this field elsewhere.

Maps are oriented in a known direction, usually with North up. Navigation requires us to define frames of reference (or coordinate sets) so that we can orient ourselves in the mapped space, and in this instance "north" specifies the map's

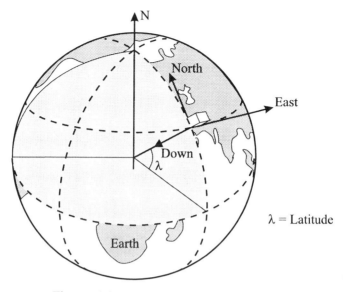

Figure 1.2. Earth-fixed axes.

reference frame. There is a frame fixed in the "fixed stars," the *inertial frame*, that Newton postulated for his laws of inertia in 1687, and inertial navigation is navigation in this frame. This is a reference frame that is independent of the motion of the vehicle.

On (or close to) the earth's surface it is more useful to work in a reference frame oriented to surrounding objects, a *local level* frame with North, East, and vertically Down as axes. In the simple (spherical homogeneous) earth local level frame, shown in Figure 1.2, Down is a vector pointing at the center of the earth at the angle of latitude λ, North is the horizontal vector in the plane of Down and the earth's spin axis (True North), and East is the horizontal normal to that plane at the observer's site. The earth rotates once in 24 hours, so its rotation rate is 15 deg/h; the horizontal and vertical earth's rate components Ω_{eh} and Ω_{ev} in local level axes are

$$\Omega_{eh} - 15 \cos \lambda$$
$$\Omega_{ev} = 15 \sin \lambda \text{ deg/h}$$

where λ = latitude.

East is the direction in which there is no earth's rate, a fact that is used to align inertial navigation systems because it is much easier to find a null than the peak of a maximum. Gyroscopic north-finding systems actually find east!

Vehicles have their own axis set, Roll, Pitch and Yaw, shown in Figure 1.3, corresponding to the conventional use of the terms.

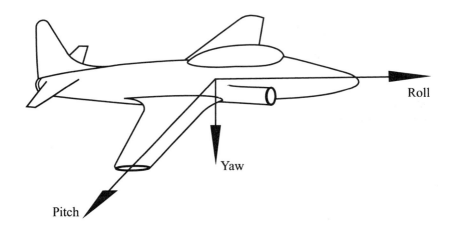

Figure 1.3. Vehicle axes set.

The Inertial Navigation Process

The inertial navigation system, INS (or unit, INU), is made from a navigation computer and a set of gyroscopes and accelerometers that measure in Newton's inertial axes, generally called *inertial sensors*. The group of inertial sensors is commonly called an *inertial measurement unit* (IMU) or an *inertial reference unit* (IRU). Once aligned to a set of reference axes (such as the "North, East, Down" set), the sensors provide distance measurements and the navigation computer carries out the continuous dead reckoning calculations.

The inertial sensors might be mounted in a set of gimbals so that they stay level and head in a fixed direction no matter how the vehicle moves. This construction is called a *navigation platform*. Alternatively, the instruments might be attached to the vehicle, in which case they measure its motion components in the vehicle axes set, and the system computes direction traveled in the reference axes by transforming the measurements from the vehicle axes to the reference axes. This is called a *strapdown* system, jargon for instruments "strapped down" to the vehicle.

To navigate inertially, we first measure the accelerations in the directions of the navigation axes, and if our instruments are not perfect, we might compensate their readings by removing biases or scale factor errors (defined in Chapter 2), perhaps known as a function of the system's measured temperature.

Second, to find the vehicle's vertical acceleration, we subtract gravity from the "Down" accelerometer output, perhaps using a gravity model to allow for the variation of gravity with latitude and longitude. Standard gravity has been defined as 9.80665 m/s². "Plumb bob" gravity defines the local vertical "Down" direction.

Its magnitude includes earth's gravitation, tidal effects (from the moon), and centripetal effects. The earth's rotation causes a centripetal acceleration that is greatest (about 3 mg) at the equator and zero at the poles. It is directed normally to the earth's spin axis and is not therefore directed along the vertical, except at the equator. In addition, the earth is not a sphere; rotation has flattened it, causing a mass concentration at the equator, so that the value of g varies with latitude. Precise navigators, like those in nuclear submarines, carry gravity gradiometers to correct for gravity vector errors; because of this unevenness in mass distribution, the vector direction is not necessarily along an earth's radius [9, 10]. In space we can use Newton's Law of Gravitation to compute the local gravity, but once on a foreign planet, we lose accuracy until we can map its gravity fields.

Third, we integrate the accelerations over a known time, once to get the velocity, twice to get the distance traveled. For periods of constant acceleration we can apply the "equations of motion" to find the distance traveled, s:

$$v = u + at$$
$$s = ut + \frac{1}{2}at^2$$

where
  a = acceleration
  v = velocity after time t
  u = initial velocity

Because integration is the process of summing the outputs at frequent, known intervals, we must know the time interval accurately as it enters as a squared term in the distance computation.

Fourth, we measure the rotation rates either from gimbal motions in a stabilized platform or directly with gyroscopes in a strapdown system. We then compensate for gyro bias and possibly scale factor errors, and we determine a new heading.

Fifth, we compensate for earth's rotation if we are in local level axes; otherwise the platform would be space stabilized and would seem to tilt in the vehicle axes set.

Finally, the combined distance and heading data give us an updated dead reckoned position to display. Then we go back to the beginning and do it all over again, until the end of the journey.

Inertial Platforms

An inertial platform uses gyros to maintain the accelerometers in a fixed attitude, i.e., the gimbaled platform serves to define the directions for the measurements of acceleration. A single-axis platform, shown in Figure 1.4, consists of a gyro mounted so that its sensing axis or *input axis* (IA) is along the axis of the platform, which is set into the vehicle in bearings. Electrical power for the gyro comes in,

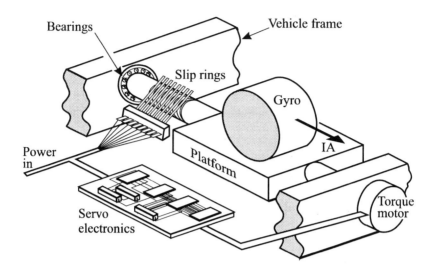

Figure 1.4. A single-axis platform.

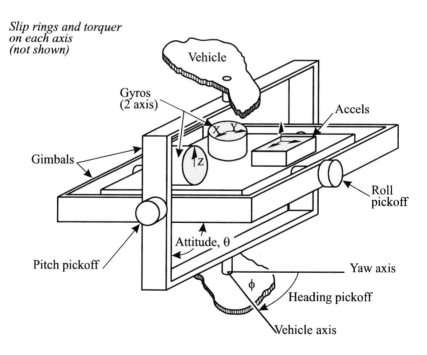

Figure 1.5. A three-axis platform.

and the gyro's output goes out, through slip rings. The platform can be driven around its axis by an electric torque motor. The gyro output, which indicates an unwanted platform rotation, drives the torque motor through a servo amplifier; it basically provides the torque to overcome the friction in the slip rings and bearings.

If the platform had no gyro and the vehicle rotated around the platform axis, the platform's inertia would tend to keep the platform aligned in space. But the platform would slowly accelerate due to the torque transmitted through the bearings and slip rings, and the inertial reference would be lost. By adding the gyro, we provide a means of sensing any platform rotation with respect to inertial space, down to the limit of the gyro's resolution. The gyro does not measure anything in the inertial platform; it is only used to maintain a fixed position (null operation), and so it does not need to be able to measure large rotation rates.

For navigation in three dimensions, we must expand our platform to three axes by adding gimbals, as shown in Figure 1.5. Here we have shown gyros with two sensing axes each (*two-degree-of-freedom* gyros), and we schematically show a group of three accelerometers. We have also provided a *pickoff* on each gimbal axis, a device that measures the angle between two gimbals; the torquers, slip rings, and three servo loops are there but are not shown. If the gimbals are aligned with the vehicle axes before setting out, then the gimbals, being gyro stabilized, provide the reference for measuring the attitude and heading of the vehicle during the journey.

The gyros alone will try to maintain the platform aligned in inertial space. If the platform is operating in local level coordinates, the navigation computer must keep the platform horizontal. It does this by sending command signals to the gyros that otherwise would fight the gimbal motion. The roll and pitch accelerometers can be used to level the platform if we know that it is not accelerating. The relationship between the sensors, the gimbals and the navigation computer is shown schematically in Figure 1.6.

Heading and Attitude Reference Systems

Rather than use an expensive inertial navigator, many aircraft use a simplified version that indicates the direction of flight (the heading) and the attitude in roll and pitch. This Heading and Attitude Reference System (HARS) is a platform with three stabilized axes and two horizontal accelerometers. The accelerometers provide a vertical reference to set the initial level before takeoff, whereas a flux valve (a kind of magnetic compass) can be used to set the magnetic heading. This heading can be corrected to True North using the local value of the magnetic variation. Alternatively, the heading can be set by gyrocompassing (page 16). Once under way, the HARS provides flight information under all visibility conditions.

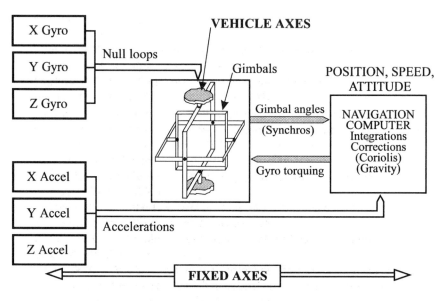

Figure 1.6. The basic platform system.

Schuler Tuning

Imagine that we have a pendulum hanging in the vehicle that we are navigating, intended to provide a vertical reference. But as we accelerate horizontally, the pendulum tilts, giving a false vertical indication. Schuler showed that this would not occur with a pendulum of 84 minutes period, and we can make a compact compound pendulum with this period using an ordinary accelerometer and a servo loop with the correct characteristics. Correcting an inertial system so that it does not tilt when accelerated is known as *Schuler tuning* it.

Gimbal Lock

Three-axis platforms have a limitation that can cause them to "lock up" during aerobatic maneuvers. This *gimbal lock* can be caused by a set of aircraft movements that cause two of the gimbal axes to become aligned, followed by a rotation in the plane of the aligned axes. How this happens is much easier to see than to read, so Figure 1.7 shows a sequence that tumbles the platform. A plane flying straight and level climbs vertically. Now the roll gimbal is aligned with the pitch gimbal, and a turn over the wing, around the aircraft yaw axis, forces the pitch gimbal over and makes the system lose alignment. A fourth gimbal [10], a duplicate roll gimbal, can be added to overcome this limitation, although it adds considerably to the system's cost and size.

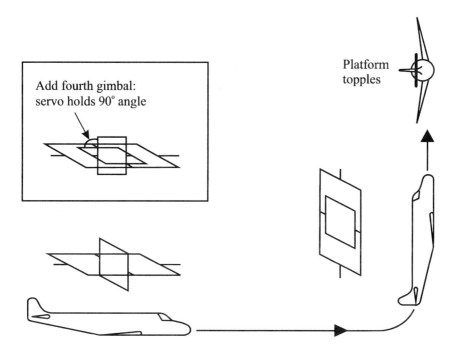

Figure 1.7. Gimbal lock.

For further reading, books by Markey and Hovorka [12] and Savant et al. [13] describe inertial systems in more detail.

Strapdown Systems

The platform system's complex mechanical construction goes against the trend in machine design. Since the 1960s, technical advances in electronics and optics have reduced their size and cost, while there has been little progress in reducing the cost of mechanisms, apart from robotic fabrication. Generally, lower costs and higher reliability have come from replacing mechanisms with electronics.

Navigation systems have benefitted from this technological culture; the strapdown system is the outcome. The strapdown system replaces gimbals with a computer that simulates their presence electronically. In the strapdown system the gyroscopes and accelerometers are rigidly mounted to the vehicle structure so that they move with the vehicle, as shown in Figure 1.8. Now, unlike the platform gyros, the strapdown gyros must measure the angles turned, up to the maximum rotation rate expected. Airplanes can experience short-term rates up to 400 deg/s,

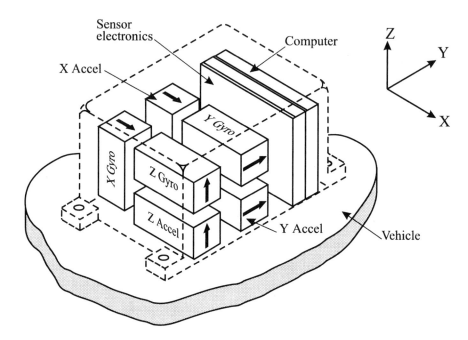

Figure 1.8. The strapdown system.

so a gyro with 0.01 deg/h performance has a dynamic range of 10^8 —two orders above the platform gyro.

As the vehicle travels, the gyros measure the yaw, pitch, and roll angles turned in a short time (say, 0.01 s) and pass them to a computer that uses them to resolve the accelerometer outputs into the navigation axes set. Simplifying to two dimensions, we can easily express the coordinate transformation between the navigation axis set (x_{nav}, y_{nav}) and the body axis set (x_{body}, y_{body}), at a time when the z-axis gyro has measured that there is an angle θ between them. The accelerometers fixed in the body axes record accelerations a_x, a_y.

Converting to navigation axes we get

$$(a_x)_{nav} = a_x \cos \theta - a_y \sin \theta \qquad (1.1)$$

$$(a_y)_{nav} = a_x \sin \theta + a_y \cos \theta$$

When adding the third (z) axis we must use more complicated transformations because the rotations are noncommutative. Typical systems use direction cosines, and the transformation between the inertial set [X_i, Y_i, Z_i] and the body axes set [X_b, Y_b, Z_b] is expressed as follows:

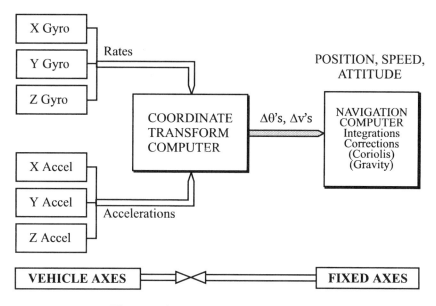

Figure 1.9. The basic strapdown system.

$$r^i = \begin{vmatrix} r_x \\ r_y \\ r_z \end{vmatrix} = C_b{}^i r^b$$

$$C_b{}^i = \begin{vmatrix} c_{11} & c_{12} & c_{13} \\ c_{21} & c_{22} & c_{23} \\ c_{31} & c_{32} & c_{33} \end{vmatrix}$$

where c_{ij} are the direction cosines between the jth axis in the inertial frame and the kth axis in the body frame; Britting [13] describes the necessary computations. The sensors and computer in a strapdown system are arranged as shown schematically in Figure 1.9.

System Alignment

The first step in inertial navigation is aligning the platform to the navigation axes or, equivalently, initializing the axes in the strapdown system's computer. We can level the platform or measure the attitude of the vehicle for a strapdown system using accelerometers. We can find North using a magnetic compass (which is not very accurate), by astronomical sightings (which means we must be able to see stars), or by *gyrocompassing*. In some cases, such as that of missiles suspended under the wing of a fighter plane, we can align the missiles' strapdown systems by

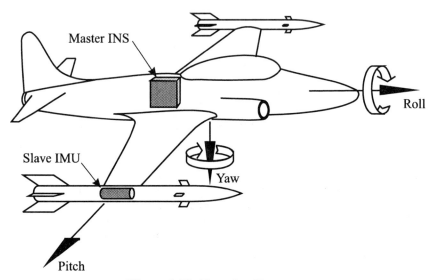

Figure 1.10. Transfer alignment.

referring their outputs to those of a master navigator in the fighter, called *transfer alignment*.

Gyrocompassing

To gyrocompass with a platform, a gyro with its input axis horizontal is connected by a servo to a platform torquer that will rotate the gyro about the vertical (yaw) axis. The servo is designed to drive the sensed rotation to zero, so the gimbal will swing around until the gyro's axis points East. But if the gyro has a bias—an output signal when there is no rotation input—it will point a little off East, enough so that the earth's rotation rate component cancels the bias. Then the platform is slewed 180° and again servoed to null. This time the bias has not changed sign but the earth's rotation component has, so, by combining the East and West readings of platform heading, both North and the gyro bias can be deduced, a big advantage for the platform system. (In passing, note that some systems rotate continuously, the "rotating azimuth" mechanization, so that instrument biases are modulated at the rotation rate and filtered out. This improves performance for given instrument errors and alignment accuracy.)

A strapdown system cannot rotate the gyros except by swinging the entire vehicle, which is usually impossible to do over the half-circle necessary for East-West gyrocompassing. All one can do is measure the three components of earth's rotation with the gyros and compute where North must be, assuming that the gyro bias has not changed since it was last measured. To find North to an accuracy of 1 mr, one must measure one-thousandth of the horizontal earth's rate. At Boston, Massachusetts, for example, HER = 10 deg/h, so for 1 mr alignment accuracy we must know the gyro bias to 0.01 deg/h, a demanding requirement met only in expensive gyros.

Transfer Alignment

Another way to align a strapdown system is to transfer alignment from a master system. A fighter aircraft can initialize wing-carried missiles, for example, by carrying out maneuvers designed to allow the fighter's master navigator to send alignment information to the missile, as shown in Figure 1.10. If the fighter flies straight and level, the master and slave velocity outputs can be matched by adjusting the accelerometer bias, and controlled rolls or turns can calibrate the slave system's gyros.

Advantages and Disadvantages of Platform Systems

Advantages

1. Simpler gyros. Because the sensor platform rotates only at the small rates needed to keep it level, the gyros need only a small dynamic range. A maximum rate of 3 deg/s would suffice for a gyro of 0.01 deg/h performance (e.g., for an aircraft navigator), a range of 10^6. Further, gyro torquer errors do not lead to attitude error. The lack of gyro rotations means that there are no anisoinertia and output axis angular acceleration errors to minimize in the design.
2. Higher accuracy. Because the accelerometer axes are always well defined, the platform navigator can be very accurate; the North and East accelerometers see no component of gravity and measure only the vehicle accelerations. The vertical accelerometer, though, measures the vehicle's vertical motion in the presence of $1g$, therefore less accurately. In an aircraft this causes altitude errors, which can be compensated with a barometer signal.
3. Self-alignment by gyrocompassing.
4. Sensor calibration by platform rotations. The other sensor biases are obtained by orienting the platform with each major axis vertical in turn, provided there is enough time.

Disadvantages

1. Complexity and cost. The gimbal structure and its bearings must be stiff so that the accelerometer axes remain defined even under vehicle vibrations, but the bearings and slip rings must have as little friction as possible. As a result, the gimbal structure is an elaborate, precisely made, expensive, mechanism.

2.   Gimbal magnetics. Each gimbal must have a pickoff and torquer. The pickoffs (synchros) measure the intergimbal angles with arc-sec resolution over a full revolution, a range of 10^6. When the system is first switched on, the torquers need to provide enough torque to accelerate the platform inertia so that it can gyrocompass and align itself to Level and North in a reasonable time. The torquers also must not leak magnetic flux, for that could upset the sensors as the gimbals move around them.

3.   Reliability. The bearings and slip rings tend to wear, degrading alignment and performance.

Advantages and Disadvantages of Strapdown Systems

Advantages

1.   Simple structure, low cost. Strapdown systems are lighter, simpler, cheaper, and easily configured for odd-shaped spaces. As it is only necessary to mount the sensors so that their sensing axes point in known directions (usually orthogonal) in the vehicle, they can be placed so that they best use the space available.

2.   Ruggedness. The simpler structure better withstands shock and vibration, and, because it is lighter, it is easier to shock mount than a platform.

3.   Reliability. There are no gimbal magnetics, no slip rings, and no bearings. The electronics that replace them are inherently more reliable.

Disadvantages

1.   Alignment. Strapdown systems are difficult to align because they cannot be easily moved. Transfer alignment is suitable for tactical systems.

2.   Sensor calibration. Again, the immobility means that the sensors cannot be calibrated in the system. Therefore, they must be stable, a burden on the sensor design. Strapdown systems rely on sensor models, using real-time compensation of inertial errors and thermal effects.

3.   Motion-induced errors. The body motions induce unique sensor errors (torquer errors, anisoinertia, output axis angular acceleration), which can be compensated to some degree.

4.   Accelerometer errors. Bias errors accumulate, and strapdown accelerometers may be subjected to components of gravity as the vehicle rolls and pitches, reducing the accuracy of the vehicle acceleration measurement and exciting cross-axis errors.

5.   The strapdown computer. Not needed in the platform system, the computer must be fast enough to do all the strapdown calculations in a few milliseconds. In a typical tactical system, a bandwidth of 100 Hz demands that

sensor compensation and coordinate transformation must be done in less than 0.01 s. This requires well-crafted program code.

The particular benefits of strapdown systems are well illustrated in the following story. One of the first strapdown systems fielded was used in the Apollo Lunar Module (LM) moon lander, as a backup to the primary platform system. The Apollo 13 spaceship had just left earth, when a fuel cell in the main ship (the service module) exploded, drastically reducing the amount of electrical power available. The ship was committed to a trajectory to the moon; it could not change direction much once in space. Mission Control decided to let the ship pass around the moon and come back to earth without landing on the moon. Because of the shortage of electrical power, the astronauts switched off the command module and used the LM. Everything possible was switched off, so that the three astronauts could stay warm enough; even so, the cabin got unpleasantly cold, down to 4°C (40°F). The platform navigator was turned off, and the crew relied entirely on the lower-power strapdown system, operating below its design temperature, for navigation around the moon and back to earth. They then left the LM and used the command module's navigator through re-entry to a safe landing [15, 16].

Aiding Inertial Navigators

Over time, inertial navigators drift from their preset alignments. Or, the initial alignment may have been corrupted by vehicle motion, with imperfect transfer of alignment and velocities to the navigator. Also, there may not have been enough time to perfect alignment. In such cases, navigators can benefit from *aiding*, from updates. There are radio navigation systems that can provide updating signals. Loran-C and Omega are land-based systems, and Transit and the Global Positioning System are space borne (orbiting satellites). Radio systems are all-weather navigation aids. Omega gives 2 to 4 nautical mile position accuracy but no velocity information, Loran-C gives about 500-m accuracy, whereas satellite systems provide better than 10-m position accuracy, and velocity data. No system provides short-term vehicle attitude data.

Alternatively, if the guidance system needs to be immune from radio jamming, it may be provided with a *star tracker* to update its alignment. This is a type of telescope with an optical detector array at its focus, which can be precisely pointed at a spot in the heavens where a star is known to be. Once flying high enough that the atmospheric turbulence, smoke, and haze do not obscure the view, the tracker locks onto the preselected star and uses the star's known position to correct for misalignment in the inertial navigator. As star trackers are expensive and require a window in the vehicle, they tend to be used only in long-range missiles, reconnaissance planes, and long-range bombers.

Satellite systems have the advantage over ground based transmitters that they are radiating almost vertically down, so that satellite signals are much less affected by

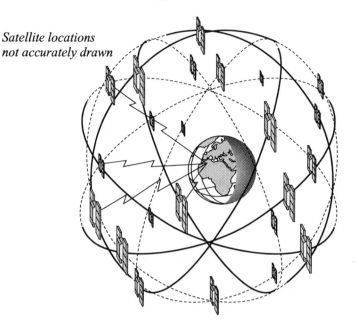

Satellite locations
not accurately drawn

Figure 1.11. The Global Positioning System satellites.

hills and the curvature of the earth. The newest United States satellite system is the Navstar Global Positioning System (GPS) and is accompanied by the similar Russian GLONASS system.

The Global Positioning System

The Navstar Global Positioning System is a space-based position and navigation system that can provide three-dimensional position to an accuracy a few meters, anywhere on or near the earth, to those with the proper receivers. It followed the Transit system, deployed in 1964 to aid nuclear submarine navigation and now being phased out.

GPS consists of 24 earth satellites orbiting at a height of 20,051 km (12,532 nautical miles) above the earth with a 12-h period (Figure 1.11). GPS was intended primarily as a military system and provides the highest accuracy (10 m or better) only to those who know its secret codes. It provides a lower-accuracy (15 to 30 m) signal to all users. The precise encrypted "P" code and the coarser clear/acquisition "C/A" signal are transmitted on separate frequencies in the L-band.

In operation, each satellite broadcasts both the P and C/A signals at precisely known times, timed by each satellite's on-board atomic clock. Each receiver has a clock, too, so it can measure the time at which it receives the signal from the satellite. As the speed at which the radio wave travels is known, the distance between the satellite and the receiver is found by multiplying that speed by the time difference between satellite transmission and user reception. Each satellite's position is known, so the user now knows position along the line of sight to the satellite. Users take data from three satellites at different angles in the sky to fix their position in three dimensions.

Now the user's clock might not be very accurate, and it is to everyone's advantage to allow it to be inexpensive, so each satellite also broadcasts the precise time in its message. As we now need a separate piece of data to decode time, we read data from a fourth satellite and compute position and accurate time. The system has been designed so that four satellites will always be visible from anywhere on the earth.

The satellites' positions can drift slightly in orbit, due to irregularities in gravity, solar pressure, and so on, so fixed ground stations observe the satellites and send orbit data to a master control center in Colorado Springs. From there, a message is sent to each satellite, telling it how much it is deviating from its predicted orbit, and providing small time corrections to its clock. The satellite incorporates these messages (*ephemeris* data) into its transmissions. All of this is naturally done in a computer in the user's receiver, and some companies make receivers small enough to carry in the hand, costing only a few hundred dollars [17, 18]. One can buy a receiver for both GPS and GLONASS [19, 20]; this makes use of the greater number of satellites in view.

The C/A code signal is called the Standard Positioning Service (SPS) and provides a positioning accuracy of 100 m (95% probability). The P code signal, the Precise Positioning Service (PPS), has an accuracy of 21 m. SPS is provided by purposely degrading the PPS by a process called Selective Availability (SA), basically by degrading the accuracy of the clock and ephemeris data.

Errors in P code signals arise from propagation in the ionosphere and troposphere. The satellite clock and ephemeris data are updated daily, so their accuracy depends on how recently they were updated. Ionospheric propagation errors are compensated (only in the PPS) by transmitting data on two frequencies and using a two-frequency receiver, while tropospheric propagation errors are modeled.

Naturally, engineers have devised ways for civilians to improve the accuracy of the SPS, using Differential GPS (DGPS). DGPS uses ground stations in well-surveyed sites, which take readings on where GPS says they are at this moment. Comparing this data with their known positions they can derive correction signals to broadcast to local receivers. Under favorable conditions, DGPS can provide better than 1 m accuracy.

There are now many more SPS users than PPS (military) users, and DGPS has circumvented SA, and GLONASS does not use SA. Therefore, people ask why not remove SA and save civilian users money? Military users would still have the benefit of the dual-frequency PPS transmission; it would still be harder to jam.

GPS allows more accurate navigation but existing charts and maps are often not accurate enough to realize the benefit [21]. However, GPS can itself be used for mapmaking and geodetic surveys, to centimeter accuracy. It is used on test ranges to check out inertial navigation systems, and it facilitates air traffic control and search and rescue missions. It is changing the inertial navigation field for all except strategic weapon guidance, for such a device is exactly what an inertial navigation system needs for a partner [22].

The INS gives very accurate attitude and distance measurements for a short time, but gyro and accelerometer errors accumulate after a while and navigation errors grow. The GPS system has exactly the opposite characteristic, for it can give accurate position fixes periodically to correct the INS, but it cannot tell you quickly if you're turning or rolling; the two together complement one another [23]. Should the vehicle lose the satellite signals for a while, perhaps because the vehicle has banked and a wing hides the antenna from the satellite or because a ground vehicle has entered a tunnel, the INS can carry on until the receiver reacquires the satellite signal. For military use the INS can keep a missile on target even if the GPS signal is obliterated by powerful jammers.

Applications of Inertial Navigation

The most accurate systems, with heading errors less than 0.001 deg/h (one revolution in 40 years!), are found in submarines and ICBMs, which need navigation with errors measured in tens of feet over long mission times. We might call these "one-mile-a-day" systems. Their accelerometers resolve gravity to a few μg while measuring vehicle accelerations of the order of $10g$, a dynamic range of 10^7.

Next, aircraft systems, with errors of less than 1 mile an hour, are found in all civil aircraft that cross the oceans, and in long-range military aircraft. While all the one-mile-a-day systems are platform systems using mechanical sensors, laser gyro strapdown systems are dominating the onr-mile-an-hour market now. Less expensive heading and attitude reference systems are used in short-range military aircraft, and, frequently, in long-range planes, to back up the INS. HARS often use mechanical gyros, but laser and fiber-optic gyro versions are appearing as they get cheaper.

Tactical weapons, such as short-range missiles, including the U.S. Harpoon, Tomahawk, and Phoenix, exclusively use strapdown systems. They currently use only mechanical sensors, but that is changing as optical gyros and micromachined accelerometers complete development. These are described in later chapters.

Conclusions

For a deeper study of inertial systems, the books by Broxmeyer [24], Draper et al. [25], Leondes [26], McClure [27], O'Donnell [28], and Parvin [29] could be consulted. The guidance of spacecraft uses special techniques, described by Hynoff [30].

GPS is having a tremendous impact on the navigation field, because GPS and strapdown systems go together well. The strapdown processor can support the

GPS, and the INS can tell the GPS receiver where to look for satellites and maintain navigation while satellites are obscured.

There is one very important exception, though, and that is the guidance of strategic weapons (including aircraft and submarines). In a war serious enough for strategic weapons to be used, one could not rely on the constellation of GPS satellites to survive. They would either be destroyed or their signals would be jammed. So strategic systems must remain self-contained (perhaps aided with a star-tracker), and the accuracy demanded of these systems is difficult to meet with strapdown gyros.

In the 1960s, when inertial navigation systems began to be used, their instruments were electromechanical and very sophisticated; they were expensive, large, and fairly fragile [31]. In the 1980s the gyros were more likely to be optical, using lasers; they were less expensive, smaller, and more rugged. In the 1990s gyros and accelerometers continue to get cheaper, smaller, and more robust, based on technologies described in later chapters of this book.

References

1.   Duncombe, R.L., R.F. Haupt, "Time and navigation," Navigation, J. Inst. Nav., 17, 4, pp. 381–386, Winter 1970–71.
2.   Dunlap, G.D, H.H. Shufelt, *Dutton's Navigation and Piloting*, United States Naval Institute, Annapolis, MD, 1970.
3.   Savet, P.H. (Ed.), *Gyroscopes: Theory and Design*, McGraw-Hill, New York, 1961.
4.   Allington, P.J.S., "The Sperry Mk 19 gyro-compass," Symposium on Gyros, Proc. Inst. Mech. Eng. (London), 1964–65, Vol. 179, 3E.
5.   Draper, C.S., "Origins of inertial navigation," AIAA J. Guidance and Control, 4, 5, pp. 449–463, Sept.–Oct. 1981.
6.   Draper, C.S., "Guidance is forever," Navigation, J. Inst. Nav., 18, 1, pp. 26–50, Spring 1971.
7.   Wrigley, W., "History of inertial navigation," Navigation, J. Inst. Nav., 24, 1, pp. 1–6, Spring 1977 (37 references).
8.   Mackenzie, D., *Inventing Accuracy,* The MIT Press, Cambridge, MA, 1993.
9.   Hildebrant, R.R., K.R. Britting, S.J. Madden, "The effects of gravitational uncertainties on the errors of inertial navigation systems," Navigation, J. Inst. Nav, 21, 4, pp. 357–363, Winter 1974–75.
10.   Paik, H.J., J-S. Leung, S.H. Morgan, J. Parker, "Global gravity survey by an orbiting gravity gradiometer," Eos, 69, 48, pp. 1601, 1610–1611, 1988.
11.   Fernandez, M., G.R. Macomber, *Inertial Guidance Engineering*, Prentice Hall International, London, 1962.
12.   Markey, W.R., J. Hovorka, *The Mechanics of Inertial Position and Heading Indication*, John Wiley and Sons, New York, 1961.
13.   Savant, C.R., Jr., R.C. Howard, C.B. Solloway, C.A. Savant, *Principles of Inertial Navigation*, McGraw-Hill, New York, 1961.

14. Britting, K.R., *Inertial Navigation Systems Analysis*, Wiley-Interscience, New York, 1971.

15. Lewis, R.S., *The Voyages of Apollo*, Quadrangle, New York, 1974. Pages 166–168 relate to Apollo 13's return.

16. Kayton, M., "Avionics for manned spacecraft," IEEE Trans. on Aerospace and Electronic Systems, 25, 6, p. 802, Nov. 1989.

17. "Collins demonstrates first hand-held Global Positioning System receiver," Aviation Week and Space Technology, 19 June 1989, p. 153.

18. Magellan hand-held GPS receiver, *Defense News*, 4 Sept. 1989, p. 21; also Aviation Week and Space Technology, 30 Oct. 1989, p. 51.

19. Eastwood, R.E., "An integrated GPS/GLONASS receiver," Navigation, J. Inst. Nav., 37, 2, pp. 141–151, Summer 1990.

20. Klass, P.J., "GPS, Glonass and Glasnost," Aviation Week and Space Technology, 5 Oct. 1987, p. 11.

21. Roeber, J.F, "Where in the world are we?" Navigation, J. Inst. Nav., 33, 4, Winter 1986–87.

22. *The Global Positioning System: A Shared National Asset*, National Academy Press, Washington, D.C., 1995.

23. Buechler, D., M. Foss, "Integration of GPS and strapdown inertial subsystems into a single unit," Navigation, J. Inst. Nav., 34, 2, pp. 140–159, Summer 1987.

24. Broxmeyer, C., *Inertial Navigation Systems*, McGraw-Hill, New York, 1964. Good for IN systems mathematical background.

25. Draper, C.S., W. Wrigley, J. Hovorka, *Inertial Guidance*, Pergamon Press, New York, 1960.

26. Leondes, C.T. (Ed.), *Guidance and Control of Aerospace Vehicles*, McGraw-Hill, New York, 1963.

27. McClure, C.L., *Theory of Inertial Guidance*, Prentice Hall, Englewood Cliffs, N.J., 1960. Mathematical treatment, covers Euler's equations, etc.

28. O'Donnell, C.F. (Ed.) *Inertial Navigation Analysis and Design*, McGraw-Hill, New York, 1964.

29. Parvin, R.H., *Inertial Navigation*, Van Nostrand, New York, 1962. Good, readable account of the basics.

30. Hynoff, E., *Guidance and Control of Spacecraft*, Holt, Reinhart and Winston, New York, 1966.

31. Slater, J.M., *Inertial Guidance Sensors*, Reinhold, New York, 1964.

2
Gyro and Accelerometer Errors and Their Consequences

Inertial navigation is an "initial value" process in which a vehicle's location is deduced by adding distances moved in known directions to the known position of the starting point. Errors in the deduced location come from imperfect knowledge of the starting conditions, from errors in the strapdown computation, and from errors in the gyros and accelerometers (referred to as *sensors*). In Chapter 1 we alluded to gyro bias when describing gyrocompassing, without defining it; here we will consider sensor performance more rigorously. For the most part we will follow the terminology used in IEEE Standard 528 [1].

The *bias* of a sensor is the signal it gives when there is no input. The errors caused by a misaligned system, or one with uncompensated instrument biases, depend on the configuration of the inertial system, be it strapdown or platform, the number and type of gyros and accelerometers, and how the gimbals are arranged. Britting [2] has provided a unified error analysis for terrestrial inertial systems and used it to compare performances under different conditions. However, it is useful to have some rules of thumb to give approximate values for system errors caused by deficient instruments or improper setup. We often find it necessary to "think on our feet" and bound a problem with such simplifications, so we will now develop a few of these rules, leaving those who need precision to refer to Britting's book.

Effect of System Heading Error

Given a misalignment α in heading, at speed v and travel time t the cross-track distance error, e_c, is

$$e_c = \alpha vt$$

For 1 mr error (about 3 arc-min) and a speed of 500 mph, this error will be 0.5 mph. In an aircraft, this same error in pitch attitude would make the plane climb or dive, which would be indicated by the barometric altimeter and corrected by the

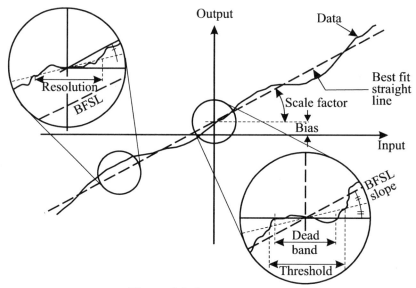

Figure 2.1. Sensor output errors.

autopilot. As 1 deg is approximately 17 mr, a gyro with a 60-deg/h drift rate allows heading to drift by 17 mr/min.

Scale Factor

An inertial sensor provides an output signal in response to either rotation or acceleration. The signal might be analog, such as a voltage proportional to the input, or digital, such as frequency or a binary number. As most systems are now controlled by digital computers, digital output is preferred, and outputs from analog sensors are digitized by A-to-D converters.

The scale factor is the ratio between a change in the output signal and the change in input. As most sensors provide an output signal that is directly proportional to the input, the scale factor is a single number, defined as the slope of the best straight line fitted by the method of least squares to the data obtained by varying the input over a specified range (Figure 2.1). The scale factor K is

$$K = S/I \qquad (2.1)$$

where
S = output signal
I = true input

The output of digital sensors is described by $1/K$, the *pulse weight*; it might have units of arc-sec/pulse. However, the inverse of the scale factor is sometimes quoted in specifications, even called "scale factor," using units of deg/h/mA, deg/h/Hz, g/Hz, and so on. In this book we will use K defined by Equation 2.1 (and IEEE STD 528) as the scale factor. Sometimes an instrument will have a different scale factor for positive and negative inputs, known as scale factor *asymmetry*.

Instruments will have at least one *input axis* defined. The IEEE standard defines the input axis (IA) as the axis along which (for an accelerometer) or about which (for a gyro) an input causes a maximum output. It is the direction about which the scale factor is defined; it is often defined as the normal to a mounting plane. For stable scale factor, the IA must remain stable over time, temperature, vibration, and shock. One can think of the scale factor as a vector, with the IA providing its direction.

Nonlinearity and Composite Error

The scale factor may not be exactly linear but may have second- or higher-order terms relating signal to input. To see these, it is usual to run a test in which the acceleration or rotation rate is varied in steps, and the sensor output is measured. The data are fitted to a straight line by the least-squares method, and the residuals are plotted. Residuals are the differences between the actual outputs at a given rate (acceleration) and the value that would be predicted using the calculated scale factor. If the residuals are randomly scattered about the axis, no further modeling is possible, but if they form curves, second- or third-order curve fits can provide a better model for the scale factor. The standard deviation of the residuals, called the *standard error*, can be used to specify instrument quality. The *composite error* is the ratio of the largest residual to the full-scale range, and it combines errors due to hysteresis and resolution, among others.

System Error from Gyro Scale Factor

Gyro scale factor errors are less important in platform systems than in strapdown systems. In platforms, the gyro input is always close to zero, and the torquer is commanded with a small signal to maintain the local level. This, with the erection and alignment system (providing the initial attitude and heading), requires no more than 0.1% accuracy. But in strapdown systems, the gyro must measure all vehicle rates—which can be as high as 400 deg/s—so the scale factor determines system accuracy directly. For example, if a mission involves an airplane circling for long periods in the same direction over a point, as it might if it were tracking a submarine, the Down axis gyro scale factor controls heading error. The heading error after time t, while circling at rate Ω, due to scale factor error $\Delta K/K$ is

$$e_h = (\Delta K/K)\Omega t$$

Let $\Omega = 1$ deg/s (one circuit in 6 min), and let $\Delta K/K = 100$ ppm. Then

$e_h = 0.36$ deg/h

Asymmetry

Small values of asymmetry, of the order of 100 ppm, are usually acceptable, except in applications where the instrument spends much of the mission oscillating about null, because then the system can rectify the asymmetry into a bias. The effective scale factor is

$$K = K^+\zeta^+ + K^-\zeta^-$$

where
   $K^+ =$ scale factor for $\Omega > 0$

   $K^- =$ scale factor for $\Omega < 0$

   $\zeta^+ = 1$ and $\zeta^- = 0$ for $\Omega > 0$

   $\zeta^+ = 0$ and $\zeta^- = 1$ for $\Omega < 0$

Bias

Generally, when there is no input, the signal is not zero; there is some offset caused by manufacturing imperfections, called the *bias* or *zero offset*. When we introduce bias B we write

$$S/K = B + I \tag{2.2}$$

For mechanical gyros, bias is sometimes called the *acceleration-insensitive drift rate*. For example, bias can be due to the torque from the mechanical null of the instrument suspension (e.g., the hinge in an accelerometer) not coinciding with the pickoff null or to magnetic contamination of a sensing element interacting with a magnetic field. If the system must quickly find North, the horizontal gyros and the leveling accelerometers must have repeatable biases from turn-on to turn-on.

System Error from Accelerometer Bias

Accelerometer errors accumulate into distance errors according to the equation of motion

$$s = ut + \tfrac{1}{2}at^2 \tag{2.3}$$

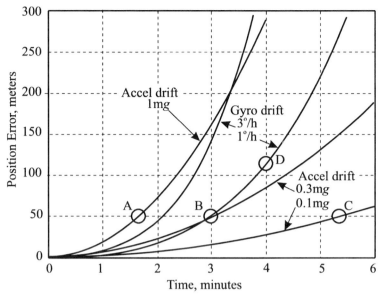

Figure 2.2. Sensor-induced position error.

where
 u = initial velocity
 a = acceleration
 s = distance traveled in time t

An uncompensated accelerometer bias of 1 mg in a journey of 1 h (starting from rest, u = 0) gives a distance error s = 65 km.

A transoceanic aircraft navigator must have much better accelerometer compensation than that to be useful. This level of performance is closer to the needs of tactical weapons navigators or of GPS-aided navigators.

Figure 2.2 shows the position errors caused by some specific accelerometer and gyro errors. Point A shows the position error of 50 m due to a 1-mg accelerometer bias error over a time of 100 s, calculated from (2.3), and points B and C show the times taken to accumulate a 50-m error with better accelerometers, with bias errors of 0.3 and 0.1 mg.

Generalizing these calculations we can see that a bias error gives a velocity error e$_v$ and a position error e$_p$

 e$_v$ = 0.589 m/s/mg/min
 e$_p$ = 17.66 m/mg/min^2

Note that a 10-mg accelerometer error causes a 6-m/s velocity error in 1 min, so external speed measuring devices like ships' logs, airspeed indicators, and Doppler radars are valuable for bounding this error source.

Tilt Misalignment

A system tilt misalignment β in roll or pitch will be indistinguishable from an accelerometer bias error, and the equivalent bias is given by

$$B_t = g\beta$$

Therefore, the position error e_t due to level tilt about North and East is

$$e_t = \tfrac{1}{2}B_t t^2 = \tfrac{1}{2}g\beta t^2$$

System Error from Accelerometer Scale Factor Error

Accelerometer scale factor error causes navigation errors only when the vehicle is accelerating (or sensing gravity, of course). Assume that for a mission the acceleration is less than $2g$ most of the time. A 1% scale factor error will give an acceleration uncertainty of 20 mg, which is indistinguishable from a bias, and from the bias calculations given earlier, the distance error is 1.3 km/h. It is clear that most missions would need better than 1% scale factor accuracy, by a factor of at least 10 .

However, for ICBM guidance, the thrust-axis accelerometer scale factor is the largest error source. This unit needs a full-scale range of $50g$ with scale factor accuracy of better than 1 ppm; for decades it has been a Pendulous Integrating Gyroscope Accelerometer described in Chapter 7.

System Error from Gyro Bias

A gyro bias is a change in angle over time, a time-varying misalignment that cross-couples into the accelerometer channel. Figure 2.3 illustrates a situation in which the z-axis gyro bias B_z causes a misalignment and couples an acceleration along x into the y-accelerometer. Then the z-gyro error angle, ϵ_z is

$$\epsilon_z = B_z t$$

and the y-accelerometer error is

$$e_y = a_x \epsilon_z = a_x B_z t$$

In this expression the gyro error angle is in radians and is assumed to be small. Integrating the y-acceleration over time, the y-velocity error is

$$e_{vy} = \tfrac{1}{2}a_x B_z t^2$$

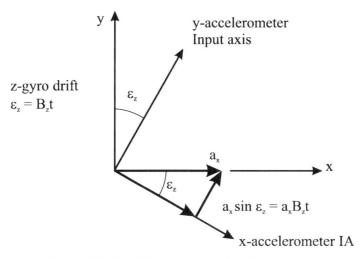

Figure 2.3. Gyro bias causes acceleration error.

Integrating again over time, the y-distance error is

$e_{sy} = a_x B_z t^3/6$

Because of the t^3 dependence, gyro drift quickly builds up distance errors. As an example let us calculate the position error caused by a 1-deg/h gyro bias error over 4 min, when $a_x = 1g$.

$e_{sy} = 9.8 \text{ m/s}^2 \times 1 \text{ deg/h} \times (4 \times 60)^3/6$ (irrational units)

Now 1 deg/h = 1 arc-sec/s, and 1 arc-sec = 5×10^{-6} radian, so

$e_{sy} = (9.8/6) \times 240^4 \times 5 \times 10^{-6} = 113$ m

This point was shown as D in Figure 2.2.

Random Drift

If the sensor is allowed to run on a stable base, sometimes referred to as a *tombstone test*, its output will wander some small amount due to disturbances inside the sensor. Such a disturbance could be ball bearing noise in a mechanical gyro; it is called *random drift* and can be characterized by the standard deviation of the output measured periodically for some specified time. Mechanical sensors can have characteristic random drifts caused by their construction; the noise may have peaks at particular frequencies. Mechanical gyros with ball bearing wheels have output noise characterized by the ball size, the number of balls, and the

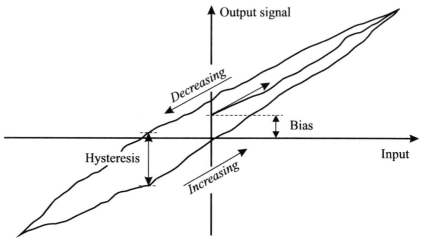

Figure 2.4. Output hysteresis.

operating speed, and by low-frequency beats due to the slight differences between the almost identical bearings at each end of the spindle.

These are usually called *in-run* drifts, to distinguish them from turn-on uncertainties. They may depend on temperature and may change as the sensor ages (particularly in ball bearing gyros).

Random Walk

Optical gyros characteristically have an approximately white noise rate output, which causes a long-term growth in angle error, called *random walk*. Their root mean square random drift will decrease as the square root of the periodic sample time, as described in Chapter 11.

Dead Band, Threshold, and Resolution

Sensors have a lower limit below which they cannot detect small changes in input, which can be regarded as a dead band around null. Perhaps they rely on movement between parts that stick together until a critical load is exceeded, colloquially known as *stiction*; mechanical sensors can suffer from stiction. Or perhaps the sensor output is noisy, and as the sensor noise output represents an uncertainty in input, small changes in input will be masked by the noise; optical gyros suffer from this deficiency.

No matter what the cause, we define the *threshold* as the largest value of the minimum input that produces an output of at least half the expected value. We call

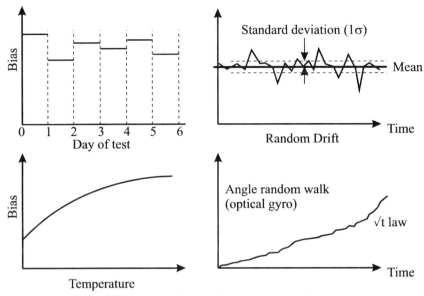

Figure 2.5. Gyro and accelerometer bias.

the minimum measurable input the *resolution*, and we define it as the largest value of the minimum input that produces an output of a specified proportion of the value expected using the scale factor. These characteristics were shown in Figure 2.1. Whereas threshold applies only around zero input, resolution affects reading accuracy at all input levels.

Hysteresis

Figure 2.4 shows a sensor's output with an exaggerated separation between the trace taken with input always increasing and that with it always decreasing. The vertical separation between the traces is due to hysteresis, and the largest separation is defined as the hysteresis error.

Day-to-Day Uncertainty

The scale factor and bias may vary from day to day by as much as 10 times more than their in-run random drifts. They may slowly change over time, perhaps from the aging of internal components such as magnets and bearings, from laser gas contamination, or from mirror or optical fiber aging. They may depend on the temperatures experienced in storage, on storage vibration and shock, on magnetic fields, and so on. Consequently a user will specify the storage and operating conditions, and the sensor supplier will be expected to demonstrate that the sensor

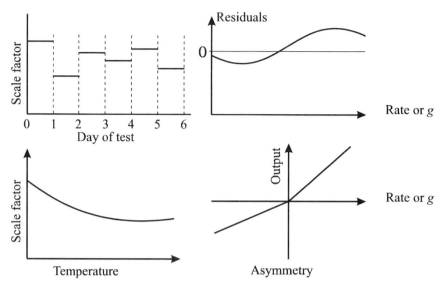

Figure 2.6. Gyro and accelerometer scale factor.

family will maintain its calibrated scale factor and bias over a long period when exposed to these conditions.

The foregoing errors are summarized graphically in Figures 2.5 and 2.6, which illustrate bias and scale factor sensitivities to time and temperature.

Gyro Acceleration Sensitivities

Gyros measure rate, and accelerometers measure acceleration. Unfortunately, gyros also sometimes respond to acceleration, and accelerometers sometimes respond to rotation. The principal gyro g-sensitivities, *mass unbalance* and *anisoelasticity*, are described in this section, and rotation sensitivities are described in the next section.

g-Sensitivity

The gyro sensitivity colloquially called "mass unbalance" is due to imperfections in fabrication caused by assembly tolerances. Mass unbalance is linear in acceleration, with units of deg/h/g, but there are other first-order g-sensitive terms that are not mass unbalances. (We will describe one, the DTG quadrature term, in Chapter 9.) Provided the value is small enough and stable, g-sensitive terms can be compensated in a navigation system, because we know the accelerations acting along the axes.

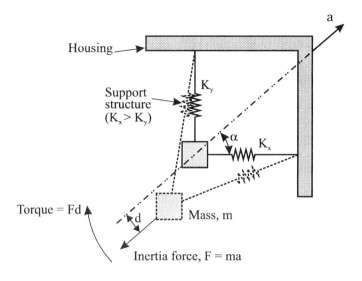

Figure 2.7. Anisoelastic torque.

Anisoelasticity

Mechanical gyros also output a signal component proportional to the square of the acceleration, called the *anisoelasticity* or the *acceleration-squared drift rate*, caused by mismatches in the stiffnesses of the members supporting the sensing element. A simple model comprising a mass supported on two springs illustrates anisoelasticity (Figure 2.7). If the model is accelerated at an oblique angle, and if the springs are not equally stiff, the mass will move off the line of action of the acceleration and there will be a torque about the normal to the plane of the figure. Referring to Figure 2.7, the torque magnitude is

$$T = \tfrac{1}{2}m^2a^2[1/K_x - 1/K_y] \sin 2\alpha \qquad (2.4)$$

where
    m = supported mass
    a = acceleration
    K_x, K_y = support stiffnesses in the x and y directions
    α = angle between the acceleration and the x-axis

Equation (2.4) has a maximum value when $\alpha = 45°$. This equation is sometimes written in terms of the spring *compliances*, the reciprocals of the stiffnesses.

    If the acceleration is a vibration at a frequency well below the system natural frequency, the displacement d (Figure 2.7) changes sign in phase with the sign

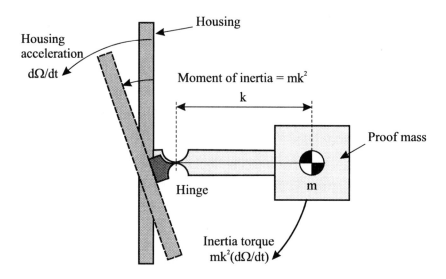

Figure 2.8. Accelerometer angular acceleration error.

change in the inertia force F, so that the torque is rectified. Assume $\alpha = 45°$, and let $a = a_0 \sin \omega t$. Then

$$T(\omega) = \tfrac{1}{2}m^2(a_0^2 \sin^2 \omega t)[1/K_x - 1/K_y]$$
which has an average value of

$$T(\omega)_{ave} = \tfrac{1}{4}m^2a_0^2[1/K_x - 1/K_y]$$

Rotation-Induced Errors

Rotation-induced errors are the angular acceleration sensitivity and the anisoinertia error. Angular acceleration can cause errors in strapdown systems, where the gyros and accelerometers experience the vehicle rotations. Mechanical sensors, being made of parts with finite mass, experience inertia forces when they are accelerated, which, if they act at some distance from a pivot, produce a torque that can be indistinguishable from a rotation or acceleration signal. This rate-induced error is caused by a net centrifugal torque arising from asymmetries in gyro or accelerometer component inertia. A rotating part tries to turn itself so that it rotates about the axis of greatest moment of inertia.

Angular Acceleration Sensitivity

Figure 2.8 shows how a pendulous accelerometer senses angular acceleration about its hinge axis. While an instrument is sensing a steady nonaccelerating rotation, it gives the correct output, but under angular acceleration it gives a transient rate or acceleration error. Defining the *pendulosity* as the product of the sensing mass (m) and the distance from the center of mass to the pivot or hinge (k)

$$p = mk \qquad\qquad (2.5)$$

we can express the angular acceleration error, e_a, as

$$e_a = (I_o/p)(d\Omega/dt) \qquad\qquad (2.6)$$
$$= (I_o/mk)(d\Omega/dt) = (mk^2/mk)(d\Omega/dt) = k(d\Omega/dt)$$

where
  I_o = moment of inertia about the output axis = mk^2
  Ω = rate of turn about the output axis

Equation (2.6) tells us that the shorter the pendulum, the better. Of course, a non-pendulous accelerometer—one in which the mass translates rather than pivots—will not have an angular acceleration sensitivity.

In an inertial measurement unit we know the rotation rates and accelerations about each axis, and we know the moments of inertia of the sensor components, so we can estimate compensations for angular acceleration error, even though the indicated angular accelerations are corrupted by the accelerations themselves.

Anisoinertia

Figure 2.9 shows a structure with unequal moments of inertia about the x- and y-axes being rotated about an axis in its plane. The centrifugal accelerations $mr_x\Omega^2$ and $mr_y\Omega^2$ are unequal, giving rise to a net torque T trying to drive θ to zero

$$T = \tfrac{1}{2}(I_x - I_y)\Omega^2 \sin 2\theta \qquad\qquad (2.7)$$

where
  I_x, I_y = moments of inertia about the x- and y-axes
  Ω = rotation rate
  θ = angle between rotation and the x-axis

Equation (2.7) is sometimes written in terms of the x and y rotation components

$$T = (I_x - I_y)\Omega_x\Omega_y$$

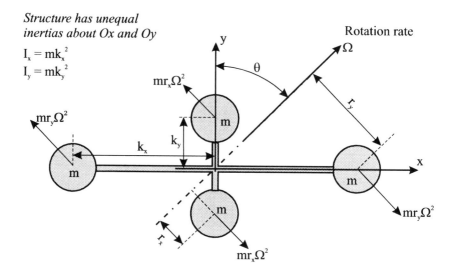

Structure has unequal inertias about Ox and Oy

$I_x = mk_x^2$
$I_y = mk_y^2$

Figure 2.9. Anisoinertia torque.

As with the anisoelasticity, the fact that Ω_x and Ω_y change sign together leads to the rectification of oscillations lower than the resonant frequency.

Angular Accelerometers

If the double integration of linear acceleration gives distance, why not use angular accelerometers instead of gyros and integrate their output twice to get the angle turned? Every few years this question arises, particularly when someone is brooding over the complexity and cost of gyros and yearning for the mechanical simplicity of an angular accelerometer [3].

One problem with angular accelerometers is that they cannot detect very low frequency or steady (DC) inputs. At high frequencies they measure well, and so they are used in autopilot stabilization systems, where vehicle attitude control comes from some other reference, and in camera and gun stabilization systems. But the absence of a low sensing threshold means that they generally cannot be used for navigation, for in situations where a vehicle makes a long, large-radius turn, or worse, starts a very slow roll, the angular acceleration goes unmeasured if it is below the sensor threshold. This means that the vehicle's heading can drift severely, at an increasing rate. The vehicle could even roll over, the roll gradually getting faster, quite undetected by the angular accelerometer! Whereas this same problem can occur if a gyro sticks so that its threshold rises, the angular velocity will be bounded, and anyway, the same friction torque will have thousands of times less effect in a gyro than in an angular accelerometer.

Angular Accelerometer Threshold Error

A vehicle using an angular accelerometer could be accelerating around an axis at a value just below threshold and the vehicle's guidance system would be oblivious to it. In a time t the angular velocity ω would build to

$$\omega = \alpha_t t \ \ \text{rad/s}$$

where α_t = threshold, rad/s². The angle error or drift is then

$$\theta = \tfrac{1}{2}\alpha_t t^2$$

So while the gyro drift angle builds linearly with time, that of the angular accelerometer builds as its square. The cross-track error e_c due to the angular accelerometer's threshold is

$$de_c = \theta v \ dt = \tfrac{1}{2}\alpha_t v t^2 \ dt$$

where v = vehicle velocity. Integrating over time,

$$e = \alpha_t v t^3/6$$

This expression can be used to judge the feasibility of using an angular accelerometer in a navigator, by substituting threshold and mission parameters.

The Statistics of Instrument Performance

Experience has shown that the characteristics of a gyro or accelerometer are not stable; they vary from turn-on to turn-on and during operation. These errors are bounded, and the *variance* describes the likelihood of a parameter falling within a given range. The day-to-day repeatability parameter might be determined by taking a set of measurements on a given instrument at the same temperature first thing each day, the overnight storage conditions having been specified. The distribution of the measurements will be described by the variance or by its square root, the standard deviation, s:

$$s^2 = \frac{\sum\limits_{i=1}^{n} (X_i - X_m)^2}{n-1}$$

where
    n = number of measurements
    X_i = value of the ith measurement
    X_m = mean value of the X_i

The data are often assumed to be normally distributed, and the standard deviation is identified with the 1σ value of the normal distribution, implying that there is a 68% chance that the bias will be within the band X'±σ at any turn-on. The 3σ value is often taken as the absolute limit of the parameter, for there is a 99.7% chance that the bias (in our example) will be within X'±3σ at any turn-on.

For economic reasons, the instrument's characteristics must be determined from a small sample of tests, so the statistics are applied to verify that the instrument is truly a member of the population with the desired performance. Thus there is reasonable certainty that it will not exhibit larger errors during its life, unless it is wearing out. This is where reliability tests are valuable, for they will characterize the sensor life or mean time between failures (MTBF), and the MTBFs for all the inertial system components can be combined to predict the time probability of system failure. The sequence of tests carried out to verify that the instrument is typical of the population is commonly known as the acceptance test procedure (ATP), and it will include tests over time, temperature, vibration, shock, magnetic fields, and anything else that might make trouble. We will describe accelerometer and gyro testing in Chapter 15.

Typical Instrument Specifications

By way of example, Table 2.1 gives specifications that describe instruments that could be used in a GPS-aided navigator or in a mid-course guidance system for a missile with a flight time of a couple of minutes. They do not apply to any particular system, nor are they complete. Typically an instrument specification will take 30 pages to cover all the operating and storage conditions, as, for example, described in the specification format for a single-axis rate gyro [4]. In a real application there would be more detail about the environment, possibly calling up a military standard [5]. Limits to some cross-coupling, anisoinertia, and higher-order terms might be specified, particularly if the chosen sensor type has unusually high response to a particular input. For example, it is necessary to specify acceleration sensitivities for mechanical gyros but not for laser gyros, and laser gyros exhibit random walk while mechanical gyros have different noise characteristics, perhaps caused by the wheel bearings.

In subsequent chapters we will combine the error terms defined in this chapter into individual error models for the different kinds of accelerometers and gyroscopes we will be considering, and then we will describe the procedures used to measure their magnitudes. While we have covered those error terms that are generally important, in special applications and with individual instrument designs other errors may become significant. Some instruments are sensitive to magnetic fields, and some are not; some generate high-frequency noise and some low. We will indicate the characteristics of each instrument type as we discuss them.

Table 2.1. Sample Inertial Sensor Specification.

Accelerometer		
Maximum acceleration	g	50
Bias stability		
Day-to-day	mg r.m.s.	1
In-run random	mg r.m.s.	0.1
Scale factor stability		
Day-to-day	%FS	0.05
Nonlinearity	ppm	100
Asymmetry	ppm	100
Cross-axis sensitivity	mg/g^2	1
IA stability	mr r.m.s.	0.5
Bandwidth	Hz	100
Gyroscope		
Maximum rate	deg/s	400
Bias stability		
Day-to-day	deg/h r.m.s.	10
In-run random	deg/h r.m.s.	1
Scale factor stability		
Day-to-day	%FS	0.05
Nonlinearity	ppm	500
Asymmetry	ppm	100
g-sensitivity	deg/h/g	1
g^2 sensitivity	deg/h/g^2	1
IA stability	mr r.m.s.	0.5
Bandwidth	Hz	100
Environments		
Operating temperature	C	-54 to $+85$
Vibration level	g r.m.s.	30
Frequency spectrum	Hz	20 to 2000
Shock level	g	300
Duration	ms	3
(Shape, half-sine)		

References

1. ANSI/IEEE STD 528-1984. Standard Inertial Sensor Terminology.
2. Britting, K.R., *Inertial Navigation Systems Analysis*, Wiley-Interscience, New York, 1971.
3. Hartzell, R.E., "A high reliability, low cost, fluid inertia angular accelerometer and its application in real time inertial coordinate frame of reference systems," DGON Symposium Gyro Technology, Stuttgart, 1985.
4. IEEE STD 292-1969. Specification Format for Single Degree of Freedom Spring Restrained Rate Gyros.
5. MIL-STD-810. Environmental Testing, Aeronautical and Associated Equipment.

3
The Principles of Accelerometers

In this chapter we consider acceleration measurement and examine the dynamic behavior of a common accelerometer, the mass-spring second-order model, describing its responses to an impulse and to a sustained periodic driving force. We will describe open- and closed-loop (servoed) instruments and the types of servos they can use. We will also mention the principles of two new accelerometers, the surface wave and fiber-optic types.

A body's mass is the quantity of matter comprising it; it does not depend on the reference frame [1]. Accelerometers operate by measuring the inertia force generated when a mass accelerates. The inertia force might deflect a spring, whose deflection can be measured; it might change the tension in a string and change its vibrating frequency; or it might generate a torque that will precess a gyro, the measure of acceleration being the gyro's precession rate (as in the pendulous integrating gyro accelerometer, see Chapter 7).

The Parts of an Accelerometer

Accelerometers are made up of at least three elements:

1. a mass, often called the "proof mass,"
2. a suspension, which locates the mass, and
3. a pickoff, which puts out a signal related to the acceleration.

It is not always obvious where these elements are. Take a piezoelectric block with two opposite faces plated, for example, and measure the voltage across the plates when it is accelerated. The mass is the block, the spring is the piezoelectric's bulk elasticity, and the pickoff is the piezoelectric effect itself. (Such instruments are used to measure vibration; they are not sensitive enough for most navigation systems.) Some instruments are servo controlled and will also need:

4. a forcer, an electric or magnetic force generator designed to oppose the inertia force, and
5. an electronic servomechanism.

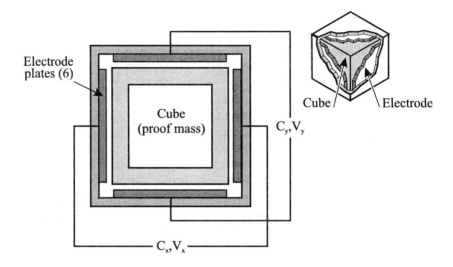

Figure 3.1. An electrostatic accelerometer.

The servomechanism, *servo* for short, is an automatic controller. A signal from mechanical motion (from the pickoff) is returned, through a feedback amplifier and the forcer, to control input conditions. Its purpose is to hold the pickoff output at null.

As an example, Figure 3.1 shows a metallic cube inside a box with six electrode plates, each spaced a small distance from a cube face. The cube may be suspended electrostatically by charging the plates with voltages V_x, V_y, and V_z. The capacitance of each plate to the cube (C_x, C_y, C_z) is a measure of the gap between the plate and the cube and serves as a pickoff. Three servos, one per pair of opposite plates, can adjust the plate voltages so that the cube is supported with equal gaps on all faces, and the voltages are a measure of the acceleration. The cube is the proof mass, and the servo-controlled electrostatic field is both the suspension and the forcer; this is a *closed-loop* sensor. A French group [2] used such an instrument to study small acceleration effects in spacecraft.

The Spring-Mass System

The spring-supported mass shown in Figure 3.2 is a basic single-degree-of-freedom accelerometer, and the relationships between the size of the proof mass, the damping, and the stiffness of the suspension determine its characteristics. We can consider the response of such a system to a force applied to the frame along the

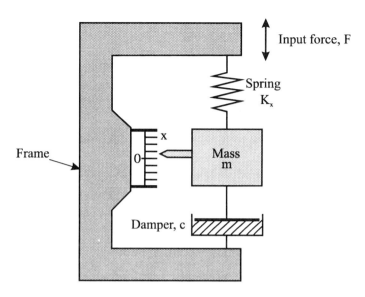

Figure 3.2. The mass-spring system.

spring's axis by summing the forces from inertia, fluid damping, and spring displacement and equating them to the applied force

$$F = m(d^2x/dt^2) + c(dx/dt) + K_x x \qquad (3.1)$$

where
  x = displacement from the rest position of the mass
  c = damping coefficient
  K_x = spring stiffness

In some instruments the damping is so high that the damping term outweighs the inertia term, so that the highest derivative in (3.1) is dx/dt and the equation is first order. It is more usual for the inertia term to dominate; in that case the equation is second-order.

If the acceleration is steady and the mass displacement is steady (any initial transient oscillation having died away)

$$m(d^2x/dt^2) = -K_x x \qquad (3.2)$$

that is, the inertia force is balanced by the opposing spring force, and x is a measure of the acceleration. The scale factor will be m/K_x. Such instruments can be purchased; the Setra Company makes one with a metal diaphragm as the spring and the capacitance between the diaphragm and the case as a pickoff.

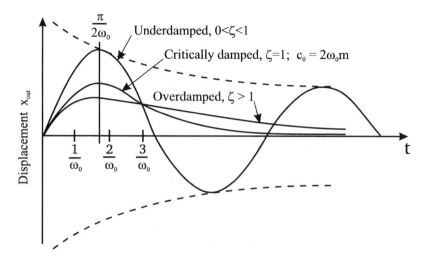

Figure 3.3. Second order system impulse response.

Accelerometers in which the displacement indicates the acceleration are *open-loop* instruments.

The solution of (3.1) is given [3] in terms of the undamped natural frequency ω_n $(= 2\pi f_n)$ and the damping ratio ζ

$$\omega_n = (K_x/m)^{1/2} \tag{3.3}$$

$$\zeta = \tfrac{1}{2}c/(K_x m)^{1/2}$$

The actual resonant frequency differs a little from f_n depending on the damping ratio

$$\omega = 2\pi f = \omega_n(1 - \zeta^2)^{1/2}$$

The impulse response of the spring-mass system, shown in Figure 3.3, describes the mass displacement over time for different damping ratios. Three curves represent the conditions:

Condition	Damping	Response
Underdamped	$0<\zeta<1$	Overshoots $x = 0$ and oscillates
Overdamped	$\zeta>1$	Sluggish, no overshoot
Critical damping	$\zeta=1$	Shortest rise time without overshoot

Note that the first peak of the underdamped response occurs at time $t = \pi/2\omega_n$. If the damping ratio is set at 0.7, there is a small overshoot but a fast response, and mechanical sensors are often designed to work at "0.7 of critical damping."

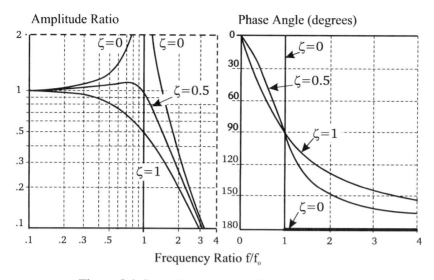

Figure 3.4. Second order system frequency response.

When the acceleration in Equation (3.1) is time varying, the pickoff output (the measure of x) varies with frequency and damping. At the resonant frequency, f_n, the output will peak. Figure 3.4 shows how the output varies in amplitude and phase with frequency, for various values of damping. When there is no damping at all, the resonant response is infinite, but this does not happen in practice as there is always some damping in a system.

Q Factor

The Q factor describes the sharpness of the resonant response of a single-degree-of-freedom, second-order, lightly damped system. Q is roughly related to ζ by

$$Q = 1/2\zeta \tag{3.4}$$

If the sharpness is represented by the linewidth Γ (the frequency difference between the two half-power points on the resonance curve), then

$$Q = f_n/\Gamma \tag{3.5}$$

Q also describes the amplitude of resonant response; it is roughly equal to the resonant amplitude divided by the driving amplitude (the transmissibility) for lightly damped systems. Finally, the Q factor relates the energy lost per vibration cycle (ΔW) to the amount stored (W), damping being the cause of energy loss:

$$Q = 2\pi W/\Delta W$$

Bandwidth

The accelerometer output should be described by a simple scale factor, constant over a useful frequency range called the *bandwidth*. Bandwidth is defined as the frequency band from zero to the frequency at which the phase of the response lags by 90° (second-order system; 45° for a first-order system).

Alternatively, the upper limit of the bandwidth can be defined by the frequency at which the amplitude ratio response is 0.5, the 3-dB point. In a true second-order system, both bandwidths will be the same. A system designer will often specify the minimum instrument natural frequency and a range within which the damping constant must lie (e.g., 0.5 to 0.9).

Open-Loop Pendulous Sensors

The simple open-loop pendulous accelerometer will have a proof mass on a spring hinge as the sensing element. A pickoff will measure the angular deflection of the sensing element, and this is the accelerometer's output. Because of this deflection, the accelerometer can suffer from cross-coupling and can rectify vibrations, the so-called *vibropendulous error*, which we will now consider.

Cross-Coupling and Vibropendulous Errors

Figure 3.5 shows a schematic pendulous accelerometer. If the pendulum is displaced by an acceleration normal to it (along the *input axis*), or if the pendulum's center of mass is offset from the hinge axis, the instrument is sensitive to accelerations along the pendulous axis. Similarly, if the hinge is twisted or misaligned in its plane, the instrument can be sensitive to acceleration along the hinge axis. Any of these sensitivities to steady-state acceleration along an axis other than the input axis is called a *cross-coupling* error.

In Figure 3.5, the pendulum is accelerated at an angle β to the input axis. Due to finite stiffness K_H the pendulum deflects a small angle α. Therefore, the deflected pendulum senses a component of acceleration along the pendulous axis, giving rise to a steady-state cross-coupling error. Also, under vibration at frequencies well below its natural frequency, the pendulum reverses its position as the acceleration changes direction (although acceleration and displacement may not be exactly in phase), so that the net torque about the hinge has a finite value, the

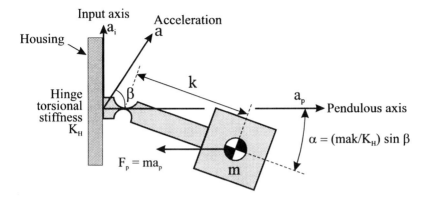

Figure 3.5. Pendulous accelerometer cross-coupling.

vibropendulous error. In Figure 3.5, the error torque about the hinge is

$$T = F_p \alpha k$$

where
 F_p = inertia force along the pendulous axis
 α = pendulum angular displacement = $(mak/K_H) \sin \beta$
 k = length from hinge to proof mass CG
 m = proof mass
 a = acceleration
 K_H = rotational stiffness of the hinge
 β = angle between acceleration and pendulous axis

Thus the steady-state cross-coupling error torque is

$$T = \tfrac{1}{2}(m^2 a^2 k^2/K_H) \sin 2\beta \tag{3.6}$$

which has a maximum value at $\beta = 45°$. Under sine vibration at frequencies well below the natural frequency, $a = a_0 \sin \omega t$, the maximum value is

$$T(\omega) = (m^2 k^2/2K_H)a_0^2 \sin^2 \omega t$$

which has an average value of

$$T(\omega)_{ave} = (m^2 k^2/4K_H)a_0^2$$

The larger K_H is (a stiff hinge) and the smaller α and k are, the lower these errors are. However, with small α, the pickoff must be extremely sensitive, and the resolution of the sensor will be set by the electrical noise in the pickoff circuit.

Pickoff Linearity

Open-loop mechanical sensors rely on the linearity of the pickoff and the spring that opposes the input force (or torque) for their scale factor linearity. Mechanical springs usually suffer from hysteresis, although springs of diamond and silicon are exceptions. Because the spring must oppose the full-scale input, it must be stiff; therefore, its mounting must be extremely stable to avoid null shifts, as very small motions represent substantial inputs.

Closed-Loop Accelerometers

We can improve instrument performance by operating in a closed loop, because then K_H in (3.6) becomes the servo stiffness and can be many orders higher than the open- loop value. The linearity problem has been transferred from the spring and pickoff to the forcer, and forcers can be made with better linearity and low hysteresis. Now the hinge spring can be made very weak, just strong enough so that it does not bend or break under shock and vibration, so its hysteresis and temperature effects are negligible. Also, pickoff linearity does not matter much, as the instrument always operates near null, although a highly nonlinear pickoff can upset the servo stability because the loop gain will change with angle. All that matters is that the mechanical and electrical null positions are stable over time and temperature.

The servo loop is intended to keep the pickoff at null all the time, regardless of acceleration, which implies a servo with "infinite" stiffness. But practical designs have infinite (very high) stiffness only at low frequencies. Practical servos have limited bandwidths by choice, because the higher the bandwidth, the higher the noise in the system; most strapdown sensors have bandwidths in the 50- to 100-Hz range. Aircraft and missile structures vibrate in flight from engine noise, air turbulence, and so on, in a band of frequencies from 20 to 2000 Hz or so. Therefore there are accelerations in the system well beyond the instrument bandwidth that cause the pendulum to deflect from null. These can lead to vibropendulous errors.

Open-Loop Versus Closed-Loop Sensors

The closed-loop mechanical sensor relies on the feedback system to restrain the sensitive element under high accelerations and rotations; therefore when there is no power to the instrument the sensitive element is unconstrained. If there is not much damping in the sensor, as in a dry pendulous accelerometer or a dynamically tuned gyro, the instrument can be damaged by nonoperating environments. This problem occurs in airplane-carried missiles, which may be carried at the wing tip for hundreds of hours, during which the airplane may make supersonic dashes and vibrate the missile severely (so-called "underwing carriage" environments). Unpowered closed-loop sensors have been known to suffer metal fatigue and fail under this treatment. But the open-loop instrument makes a virtue of necessity; it is designed so that its sensitive element reaches full-scale deflection under the maximum input, therefore it is inherently more rugged and can withstand shock and vibration whether or not it is powered.

Generally speaking, open-loop mechanical instruments are satisfactory where dynamic range does not exceed 5000 to 1 and where scale factor error can be 0.1%. Where higher performance is needed, it is better to use feedback systems to null any sensing element motion, and the output is the feedback required to maintain the instrument at null. But note that the vibrating beam accelerometer and ring laser gyro are open-loop instruments with very good performance, as we shall see in later chapters.

Sensor Rebalance Servos

Accelerometer servos share some common characteristics with gyro servos, so we can take a general look at servos relevant to both sensors. Servo design is a science usually practiced by a specialist, so we will limit this look to an overview. Oppenheim et al. [3] provide an introduction to linear feedback systems, and there are many other good books on servo design. We will consider gyro and accelerometer loops from the instrument designer's point of view, concerning ourselves with the demands that the rebalance loop makes of the sensor.

Servos may be analog or digital. Analog servos continuously vary the feedback signal to a forcer (for an accelerometer) or a torquer (for a gyro) so as to null the sensor, and the output is usually a voltage proportional to input. Digital servos pulse the signal (often a fixed current) to the forcer, and the output is the count of the number of pulses per unit time. Analog servos require the instrument scale factor to be well known over the entire operating range, whereas digital schemes operate at only one level, reducing demands on the instrument scale factor.

Binary Feedback

Binary servos provide pulses having only two states, one corresponding to full-scale positive input and the other to full-scale negative input. With no sensor input, positive and negative pulses alternate and no net force is generated. In Figure 3.6 each pulse represents either a "quantum" of velocity (if an accelerometer) or of angle (if a gyro). The measure of this quantization, the *pulse weight*, may be in arc-sec/pulse for a gyro or meter/pulse for an accelerometer. The sensor output is the difference between the number of "up" pulses and "down" pulses in a sample period.

Assume, for example, that the servo delivers current pulses. Each pulse has finite rise and fall times due to the circuit time constants, and these time constants are likely to vary with temperature. Therefore the pulse weight can vary with temperature, which is minimized by making the duration of the peak (constant)

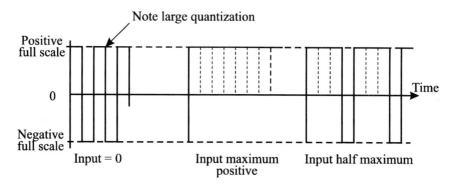

Figure 3.6. Binary feedback signals.

current large compared with the rise and fall times. This tends to give the pure binary system coarse quantization (large pulse weight).

To avoid this problem one can use a width-modulated binary loop. As in the simple binary loop, the current has only two states (plus and minus full scale) but the switching point between the states can be varied in small steps, as shown in Figure 3.7. Thus the quantization—these small steps—is fine and the pulse weight is small.

Binary loops reverse the drive current with a bridge of four switches, shown in Figure 3.8. This is called an "H-switch." The switching speed of the output transistors sets the upper limit to the quantization in width-modulated loops. If the switches have high leakage currents they will contribute bias errors to the sensor loop, as current will flow in the forcer even when the switch is "off." Timing errors, which cause the switch openings and closings to be staggered, allow current to bypass the forcer and give bias errors. If the current regulator is unstable, the scale factor will be unstable. However, with good engineering these error can be limited to 10 ppm.

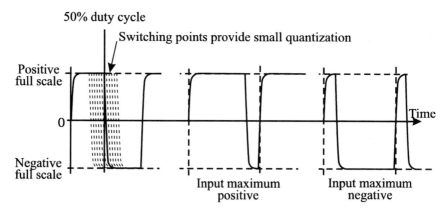

Figure 3.7. Width modulated binary feedback signals.

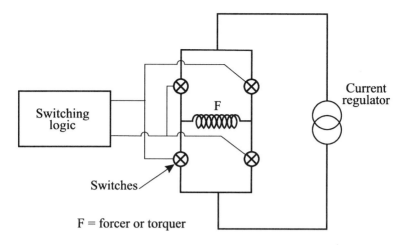

Figure 3.8. Binary switching (H-switch).

Either binary system applies the full-scale rebalance power to the sensor all the time, which has the advantage that the sensor heat dissipation is constant and therefore its temperature gradients are steady. Whereas this improves the performance of temperature-controlled instruments, it is of little value for other (unheated) tactical sensors. It can be a disadvantage in the dynamically tuned gyro (DTG), where it is difficult to dissipate heat from the torquing coils. One way around this is to use multirange loops where the peak current is switched up as the input rises. Multirange loops are more expensive because they add the hardware cost of the level switching logic, the more complex current controller, and the cost of the system software needed to keep track of the changes in scale factor.

Ternary Feedback

The analog servo supplies only the current necessary to rebalance the input, so it dissipates much lower power than the binary schemes. However, as noted earlier, it demands better scale factor linearity in the sensor. The ternary loop offers a compromise; it supplies fixed-height pulses upon demand. When there is no input, there are no pulses, as shown in Figure 3.9, so the power is not constant. Width modulation is possible to avoid large quantization, mechanized by growing the pulse in width as more torque or force is called for.

Pulse Feedback and Sensors

Pulse feedback hits the sensor with an impulse, so the pulse repetition rate must not match a sensor mechanical resonance or a nutation frequency, or the sensor will

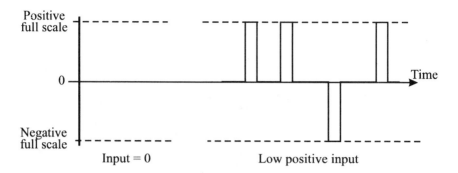

Figure 3.9. Ternary (pulse-on-demand) feedback signals.

become unstable. Even nonresonant motion can become large if mechanical or electronic damping is not provided, so that designing servos for the DTG is difficult because the DTG must have as little mechanical damping as possible (see Chapter 9).

Permanent magnet forcers or torquers must have as low a resistance (R) as possible to reduce Joulean heating (i^2R) and as low an inductance (L) as possible to keep the time constant (L/R) low. This means that the number of turns on the forcer/torquer coil must be kept low. Tuning the circuit by adding a capacitor across the coil can reduce the effective time constant. However, in width-modulated systems the tuning capacitor can cause nonlinearity errors because it acts as a bypass to the higher-frequency Fourier harmonics that arise as the pulse width varies from 50% duty cycle.

The Voltage Reference Problem

The current that rebalances the mechanical sensor is measured by passing it through a high-quality resistor of precisely known value and comparing the voltage across the resistor to a standard. This standard reference voltage ultimately determines the scale factor, and therefore the navigation accuracy. Providing a very precise voltage reference, unaffected by nuclear radiation, has turned out to be a challenging part of accurate navigation system design and is a first-order problem for force-feedback sensors.

Novel Accelerometer Principles

We have discussed the general principles of accelerometers using the pendulous sensor as an example. Other phenomena are being applied to accelerometry, notably the vibrating beam (which we will cover in detail in Chapter 5), and two new technologies offer promise, the surface acoustic wave and optical fiber sensors, so their principles are worth noting.

Surface Acoustic Wave Accelerometer

The surface acoustic wave (SAW) accelerometer uses very high-frequency (about 100 MHz) surface acoustic waves to measure the deflection of the hinge supporting a proof mass. There are different ways of using the SAW, described, for example, in [4,5]. One approach uses an acoustic interferometer. Surface waves travel on both sides of the hinge, axially along it, launched in a piezoelectric material on each side of the hinge. After a centimeter or so, the waves are picked up by another transducer, and the phases compared on the two sides. When the hinge is not bent, the two paths are equal and there is no phase difference, but when the hinge bends, the path on the outside of the bend is longer than that on the inside and there is a phase difference between the two waves. By summing the waves, one gets an intensity signal depending on the bend curvature.

This sensor is analogous to the interferometric optical gyro, described in Chapter 12, except that the acoustic wave travels with the hinge medium, whereas optical waves travel in inertial space.

Fiber-Optic Accelerometers

Fiber-optic accelerometers are sensors that use guided wave optical effects to measure the displacement of a proof mass against an elastic support. They can be connected to the indicator by optical fibers, which might be useful in situations where it is dangerous to bring electricity to the point where the acceleration is to be sensed, in fuel tanks, for example.

The optical effects used may be microbend losses, interference, or photoelasticity. Microbends are bends in the fiber whose radii approximate the fiber diameter; they cause the fiber loss to jump and therefore reduce the light transmitted. Interference between two optical beams gives a sensitive measure of their path difference, just as in the SAW accelerometer. In the Mach-Zehnder interferometer, shown in Figure 3.10, a light beam is split into two paths (A and B) that travel separately for a short distance and then recombine. If the divided beams undergo a relative phase change (perhaps due to one path being longer or passing through a medium of different refractive index), they will interfere when recombined and the intensity will be reduced. Should the phase difference on recombination be a multiple of 2π, the two waves will cancel one another and the intensity will be zero. The Mach-Zehnder is analogous to the electrical bridge circuit and can be made from two pieces of fiber separated by fused couplers. A mass attached to one fiber of such an interferometer will strain when accelerated and will give a measure of acceleration. By attaching one mass to both fibers so

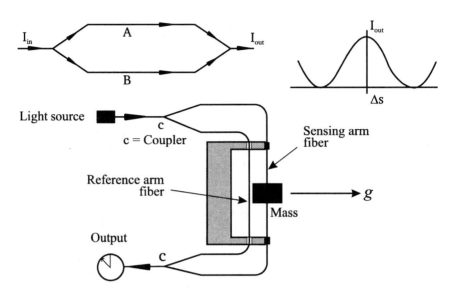

Figure 3.10. The Mach-Zehnder accelerometer principle.

that it increases the tension in one at the same time as it reduces the tension in the other gives push-pull operation with better sensitivity. Resolution of 1 µg has been reported [6]. The presence of the common mode path reduces thermal and acoustic sensitivity by many orders.

References

1. Okun, L.B., "The concept of mass," Physics Today, pp. 31–36, June 1989.
2. Bernard, A., B. Foulon, G.M. Le Clerc, "Three axis electrostatic accelerometer," DGON Symposium Gyro Technology, Stuttgart, 1985.
3. Oppenheim, A.V., A.S. Willsky, I.T. Young, *Signals and Systems*, Prentice Hall, Englewood Cliffs, N.J., 1983.
4. Hartemann, P., P.-L. Meunier, A. Jacobelli, "Elastic surface wave accelerometers," U.S. Patent 4 515 016, 7 May 1985.
5. Bower, D., M. Cracknell, A. Harrison, "A high linearity SAW accelerometer," Proc. IEEE 41st Annual Frequency Control Symposium, pp. 544–547, 1987.
6. Tveten, A.B., A. Dandridge, C.M. Davis, T.G. Giallorenzi, "Fiber optic accelerometer," Electronic Letters, 16, 22, pp. 854–856, 23 Oct. 1980.

4
The Pendulous Accelerometer

The pendulous accelerometer, one with an unconstrained single-degree-of-freedom pendulum operated in a closed loop, may well be the most common navigation accelerometer. In this chapter we will consider three types of pendulous accelerometer:

1. a generic pendulous instrument,
2. the "Q-Flex" design, and
3. the silicon micromachined accelerometer.

A Generic Pendulous Accelerometer

A generic closed-loop pendulous accelerometer might be constructed as shown in Figure 4.1. This accelerometer consists of a proof mass, a hinge, some damping, a pickoff, a forcer, and a servo loop. The three principal accelerometer axes—the input, output (hinge), and pendulous axes (IA, OA, and PA)—are also defined in the figure.

Mass and Pendulum Length

The pendulosity [the product of proof mass and the pendulum length; see Equation (2.5)] acts as the scaling parameter between acceleration and torque. Under acceleration along the IA, the OA torque is

$$T = Fk = mak = ap$$

where
  F = inertia force
  k = length from hinge to proof mass CG
  m = proof mass
  a = acceleration
  p = pendulosity = mk

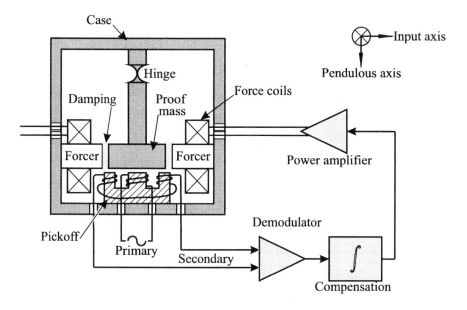

Figure 4.1. The basic pendulous accelerometer.

Therefore,

$$a = T/p \qquad (4.1)$$

In an analog instrument, the torque might be proportional to the feedback current, i

$$T = k_F i$$

where k_F = forcer scale factor. Then, $a = k_F i/p$.

Scale Factor

The instrument scale factor, which was given in Equation (2.1), is

$$
\begin{aligned}
K &= \text{output signal/input} \\
K &= i/a = p/k_F
\end{aligned}
\qquad (4.2)
$$

For high scale factor and sensitivity we require high pendulosity, that is, large m and k. But the mass should not be so large that under shock loads it creates stress levels that damage the hinge. Mission dynamics affect the choice of pendulum length. In strapdown systems, k should be small to minimize angular acceleration error [Equation (2.6)] and anisoinertia error [Equation (2.7)].

Because the scale factor is proportional to the pendulosity [Equation (4.2)], for a stable scale factor it might be necessary to compensate for the pendulum's thermal expansion. Also, pendulum length can change dynamically under acceleration along the pendulum axis; the structure can be shortened or stretched. Finally, forcer scale factor k_F may depend on the stability of its magnet's field (discussed later).

The Hinge

The hinge supports the mass and must not be overstressed under the shocks and vibrations it may receive in both storage and use. Yet it must have as low a rotational stiffness as possible about the OA, so that it does not restrain the pendulum in its response to the acceleration it is sensing. But, it must have high stiffness about the IA and PA. The hinge must not show torque hysteresis when displaced to its limit stops; that is, it must repeat its null position precisely, otherwise the bias will be unstable. It must not buckle, twist, or bend in response to accelerations either along the PA or the OA. The hinge must lie in the plane containing the proof mass center of gravity, so that accelerations along the PA do not produce cross-coupling errors [Equation (3.6)].

In metal hinges the metal composition, its crystal grain structure, and heat treatment are all carefully controlled during manufacture. Hinge geometry must be stable over temperature. In assembly, the hinge must not be bent; the mechanical null must coincide with the pickoff electrical null (to avoid a bias torque) and both nulls must be stable over temperature and time.

The Pickoff

The pickoff shown in Figure 4.1 is a form of differential transformer, in which there are two secondary windings excited by a common primary carrying an alternating current. The magnetic fluxes linking the two secondaries depend on the position of a movable armature, shown here as the entire proof mass, made of a magnetically permeable material. When the armature (proof mass) is centered, the fluxes linking the two secondaries are the same and generate the same voltage; the secondaries are connected in opposition so that there is no output voltage. When the mass moves sideways, one secondary voltage rises and the other falls, and their difference voltage increases. The increase is proportional to the displacement, and the phase of the output with respect to the primary excitation indicates the sense of the movement. Demodulating the output with respect to the primary current provides a DC voltage whose sign represents the sense of the motion.

The Forcer and Servo

The pickoff output passes to a servo amplifier, which generates a current to drive the forcer system. Figure 4.1 shows one approach in which two force coils, on either side, pull on the proof mass and null out the inertia force. The coils are driven from two amplifiers in the servo. However, this is not good design practice, because small dissimilarities between the two forcers would make the scale factor asymmetric. Also, if the forcers were electromagnets, as drawn in the figure, their forces would go as the square of the current, so that the user would have to take the square root of the current output to get the acceleration. While this is possible (and has some plausible advantages), it is usually not done because the permanent magnet forcer, described later, has lower eddy current and hysteresis losses and better linearity.

If the proof mass is magnetic, it must be shielded so that magnetic fields outside the instrument do not attract it and cause a bias. Magnetic fields in platform systems can come from the platform torquers, and the fields experienced at the instruments can vary with vehicle attitude and heading as the gimbals change direction relative to the instruments (Chapter 1).

The IEEE Model Equations

The IEEE model equations [1,2] for a pendulous accelerometer describe the deviations from the ideal output in terms of some instrument coefficients. Remembering that we are trying to measure a_i (the acceleration along the IA) in the presence of instrument imperfections, the model says that the indicated acceleration A is

$A = E/K_1$ E = output in instrument units (V, Hz)
     K_1 = scale factor, units/g
$= K_0$ K_0 = bias, g
$+ a_i$ a_i = acceleration along IA
$+ K_2 a_i^2$ K_2 = nonlinearity, g/g^2
$+ K_3 a_i^3$ K_3 = nonlinearity, g/g^3
$+ d_o a_p$ d_o = misalignment of IA to OA
     a_p = acceleration along PA
$+ K_{ip} a_i a_p$ K_{ip} = cross-coupling, g/g^2
$- d_p a_o$ d_p = misalignment of IA to PA
     a_o = acceleration along OA
$+ K_{io} a_i a_o$ K_{io} = cross-coupling, g/g^2

The scale factor is represented by the three terms K_1, K_2, and K_3, modeling up to a third-order fit, which was found by experience to be reasonable for a pendulous accelerometer.

Cross-coupling coefficient K_{ip} models the effect of the center of mass being displaced from the PA; the units for cross coupling are (g/g)cross $g = g/g^2$. The twisting of the flexure when it is loaded sideways, that is, torque produced when an acceleration acts along the hinge, can give rise to K_{io}.

When the instrument is pulse torqued, scale factor asymmetry becomes more important and a supplementary IEEE model equation is used

$$A = \frac{E}{K_1^+\zeta^+ + K_1^-\zeta^-} = K_0 + a_i + d_o a_p - d_p a_o \ldots$$

where

K_1^+ = scale factor for $a_i > 0$

K_1^- = scale factor for $a_i < 0$

$\zeta^+ = 1$ and $\zeta^- = 0$ for $a_i > 0$

$\zeta^+ = 0$ and $\zeta^- = 1$ for $a_i < 0$

An accelerometer for a strapdown system used in a dynamic environment will have angular acceleration and anisoinertia errors modeled by

$$M = I_o \alpha_o + (I_p - I_i)\Omega_i \Omega_p$$

where

I_i, I_o, I_p = moments of inertia about the IA, OA, and PA

α_o = angular acceleration about the OA

$\Omega_{i,p}$ = angular velocities about the IA and PA

The acceleration error is M/p. In practice, coefficients K_0 and K_1, the bias and scale factor, may vary from day-to-day and in-run, and they would be tested for each instrument over the operating temperature range, typically -50 to $+90°C$. The other coefficients would most likely be sampled occasionally once the accelerometer has been qualified for a program.

The "Q-Flex" Accelerometer

We can best illustrate the important design features of a pendulous accelerometer by studying the "Q-Flex" [3], a successful design by Jacobs, of Sundstrand Corporation (now part of AlliedSignal). Figure 4.2 shows the Q-Flex's construction schematically. The one-piece hinge and pendulum structure is made from fused quartz, a very stable nonconducting material. Because quartz has a thermal expansion coefficient of $0.6 \times 10^{-6}/°C$ compared with steel at $12 \times 10^{-6}/°C$, the pendulosity does not change much with temperature, and that contribution to scale

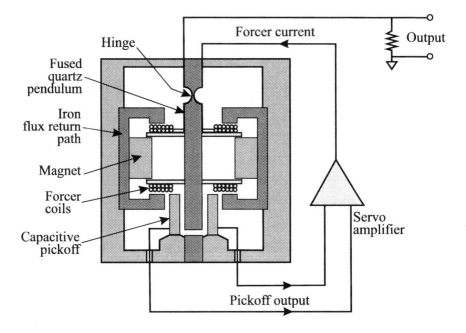

Figure 4.2. The Q-Flex accelerometer.

factor error is small. An annular cut, made almost all around, leaves two bridges of quartz that are thinned down (by etching) to form the hinge, as shown in Figure 4.3. The two-part hinge provides high resistance to angular motion in the plane of the structure (about IA) and to accelerations across the sensitive axis (along OA), minimizing cross-coupling error K_{io}.

Figure 4.4 shows an exploded view of the instrument. Two coils, mounted on opposite sides of the pendulum, form the moving part of the forcer; they carry a current that interacts with the magnetic field from case-fixed permanent magnets in the same way as a loudspeaker driver. The coils are a part of the proof mass; therefore, their mechanical stability contributes to the scale factor stability by virtue of their contribution to pendulosity.

The Capacitive Pickoff

Part of the pendulum is metallized and set between two case-fixed plates; together they form a capacitive pickoff. When the pendulum is centered, the two capacitances are equal, and when an acceleration displaces the pendulum, one capacitance increases while the other decreases. Making the capacitors part of a bridge circuit provides an output signal to drive the servo, and the servo generates a current that passes into the forcer coils by way of metallized paths across the hinge. The capacitor plates carry a low voltage to minimize parasitic electrostatic forces. The

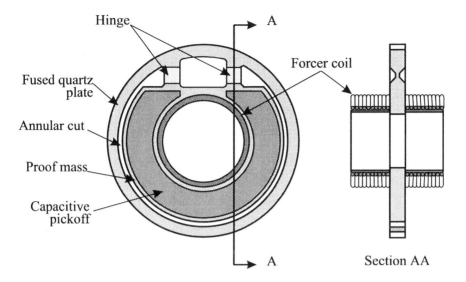

Figure 4.3. Q-Flex pendulum construction.

small clearances between the pickoff plates provide damping from the compression of the gas there, known as *squeeze film* damping.

The Forcer

Because the forcer coils carry current, they dissipate heat (at least until the day when we can use high-temperature superconducting material!). The problem with generating heat on the pendulum is that it is difficult for it to dissipate: the higher the thermal impedance, the higher the temperature in the coils. Heat can be conducted across the hinge, but that is a high-resistance path, leaving gas conduction as the remaining path. The copper wire in the coils has an expansion coefficient of $8.9 \times 10^{-6}/°C$, much larger than that of the quartz, so high coil temperatures must be avoided.

The forcer must be placed so that when it opposes an acceleration-induced inertia force, there is no net force on the hinge. This condition is satisfied when the forcer line of action is through the *center of percussion*. In the Q-Flex, this is the pendulum center of mass.

There is an optimum forcer design, one with the highest force per watt dissipated, which can be explained by examining the parameters controlling force and power. The force is directly proportional to the magnetic flux density from the magnets, the diameter of the coil, and its number of turns

$$\text{Force} = \pi BiNd \tag{4.3}$$

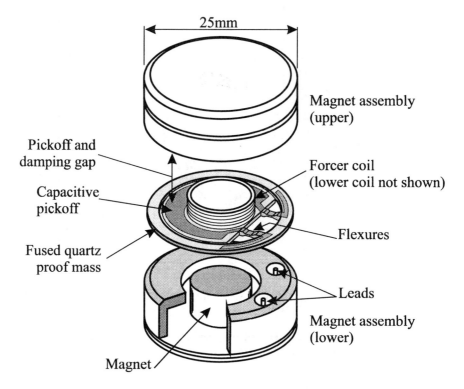

Figure 4.4. Q-Flex construction.

where
   B = magnet flux density
   i = current
   N = number of turns
   d = coil diameter

(The scale factor temperature coefficient is largely due to the change of B with temperature, a characteristic of the magnet material; temperature sensors are attached to the magnets so that compensation can be carried out in the navigation system.)
   The power dissipated is

$$\text{Power} = i^2R \tag{4.4}$$

where R = coil resistance.
   The coil resistance is proportional to the total length of the wire (that is, to N and d) and inversely proportional to the wire cross-sectional area

$$R = \rho N\pi d/(\tfrac{1}{4}\pi d_w^2) = 4\rho Nd/d_w^2 \tag{4.5}$$

where
    ρ = wire material resistivity
    d_w = wire diameter

The forcer designer must allow for the variation of ρ with temperature (the *temperature coefficient of resistance*), because the higher the temperature, the higher the resistance and the greater the power dissipated, as the forcer operates with a constant current for a constant acceleration. This could lead to thermal runaway, destroying the coil. (Copper has a temperature coefficient of $0.004/\,^\circ C$ at $20\,^\circ C$).

The magnet flux density, B, depends on the air gap in which the coil sits; the bigger the gap, the lower the flux density for a given magnet length. The relationship between B and the air gap depends on the properties of the magnet material (its *demagnetization curve*), the magnet's length-to-diameter ratio, and the shape of the magnetic circuit. Magnetics designers use empirical rules for calculating B for a given gap.

Now magnet length is restrained by the allowable size of the whole accelerometer, so we have to select a coil size that will give enough force without having too high a resistance. (Also, if we are going to pulse rebalance the unit we must keep the inductance low, and inductance goes as N^2.)

For a fixed instrument size there is an optimum air gap. If the gap is small, the flux density is large but there is little room for the coil. Conversely, a large gap will have a low flux density but a big coil. Obviously, an infinitesimal gap will give a coil with enormous resistance and high power dissipation, while the largest gap will have such a low flux density that an enormous current must flow, again generating too much heat. Somewhere in between will be the best operating point, the design with the largest force per unit power.

The forcer designer combines Equations (4.3)–(4.5), and the empirical magnet design rules in a computer program and determines the maximum force per watt. If the instrument is for digital use, the designer uses other empirical rules to calculate the coil inductance.

It is important to conduct power to the coils without causing disturbing torques on the pendulum. In the Q-Flex the current passes across the hinge through a thin metal conducting layer. This must be carefully controlled, because if the metal thickness is different on each side of the hinge, the different expansion coefficients of the quartz and the metal will generate a torque and the instrument bias will vary with temperature.

Finally, great care must be taken to ensure that the pendulum and the coil wire are free of ferromagnetic contamination. Iron inclusions or surface material from machining tools will cause large biases. Even deposited films have been found to be magnetic if not properly formed.

Other Electromagnetic Pendulous Accelerometers

While Sundstrand has been successful with a quartz hinge, other makers of pendulous accelerometers such as Bell Aerospace [4], Astronautics Kearfott, Litton of Germany (Litef) [5], Japan's JAE [6], the Israeli firm Tamam [7], and the French company Septa [8]) have used metal hinges successfully. Systron Donner has achieved inertial performance using pivot-and-jewel bearings with patented diamond torsional pivots instead of a flexure (Model 4310 is typical). Smithson's design review [9] is worth reading.

Litton, Litef, and others have used optical pickoffs in which a projection on the pendulum comes between a light emitting diode (LED) and a divided silicon photo-detector. The pendulum's shadow falls on the detector such that, at null, equal voltages come from the two halves of the detector. When the pendulum moves, the shadow moves and one half-detector voltage rises while the other falls. The difference voltage and its polarity drive the servo and forcer to restore null. The LED current can carry a superimposed alternating current to modulate the intensity, and the detector output is synchronously demodulated for greater sensitivity. The LED intensity can be controlled to be constant over temperature by a servo that varies the LED drive current in response to the sum of the dual detector chip outputs [10].

It is common to carry power to the forcer coils through very small flexleads or thin ribbons. Whereas these work very well in benign environments, they have difficulty surviving rough conditions; the ribbons bend or even break.

Moving Magnet Forcers

Why not solve the problem of carrying current to the coils over the hinge and make a more stable structure by putting the magnets on the pendulum and the coils on the case where they can dump heat easily? Figure 4.1 would illustrate this if the armature was a permanent magnet. One problem is that any magnetic fields leaking into the instrument (perhaps only from the earth's field) interact with the suspended magnet field and cause unstable biases and hysteresis. The magnetic shielding would have to be very good to avoid this problem.

Electrostatic Forcers

We have already mentioned the electrostatic forces that occur between the plates of capacitive pickoffs, which therefore must be operated at low voltage and high frequency to avoid generating error torques. However, we can use this electrostatic force to make a forcer, although they usually require high voltages (50 V and up). One advantage of electrostatic forcers is that they dissipate little power.

Generally, for electric machines a few centimeters in size, it is easier to provide large forces magnetically than electrostatically. That is why industrial and domestic

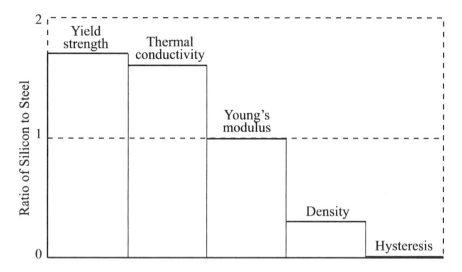

Figure 4.5. Silicon's mechanical properties.

motors are electromagnetic. But when size drops below 1 mm, the scaling laws favor electrostatic fields. It is easy to generate high voltage gradients in a small space, because the metallurgy needed to make high-flux-density tiny magnets is much less developed. This effect has led to the micromachined silicon accelerometer.

The Silicon Accelerometer

The microelectronics field has made pure single crystal silicon readily available, and silicon has excellent mechanical properties [11]. Silicon forms the same type of crystal as diamond, it is harder than most metals and has higher elastic limits than steel in both tension and compression; the two are compared in Figure 4.5. Silicon is brittle and can be cleaved; it has negligibly small hysteresis and virtually infinite fatigue life. By using anisotropic etching processes, it can be made into microscopically small mechanical devices [12,13], including small accelerometers [14]. Suppliers annually provide millions of open-loop, 1%, silicon accelerometers for use in automobile antilock brakes, air bag triggering, and so on. Here we will only consider closed-loop guidance and navigation grade instruments.

There are different designs of silicon accelerometers. The simple pendulum design works well, but others (Endevco, for one) use a diaphragm, and a few other designs have received patents [15–18]. Because this field is new, there are claims and counterclaims as to the best design; some designers use capacitive pickoffs, some use piezoresistors formed in the silicon. There are those who espouse electrostatic forcers and those who are developing very small permanent magnet

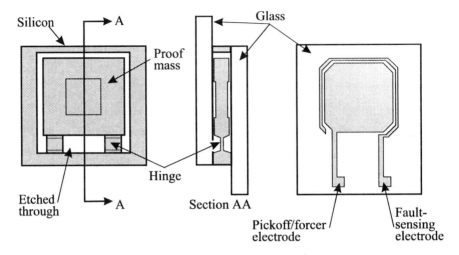

Figure 4.6. Silicon accelerometer construction.

types. Only time will tell how each design performs and what it costs. As is usual in inertial instruments, the market will most probably segment, using a different technology for different performance ranges.

A version of Northrop-Grumman's silicon accelerometer [19, 20], shown in Figure 4.6, is only 8 × 4 × 2 mm in size. (Crouzet has a similar instrument [21].) It is electrostatically forced. The pendulum is formed by etching through a silicon wafer along three sides of a square and then etching from each side partway through to form the hinge a few micrometers thick. The pendulum mass is etched down a few micrometers to form the working clearance. By anodically bonding

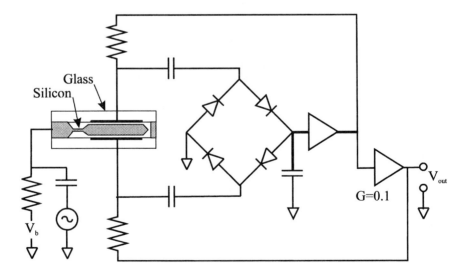

Figure 4.7. Simplified silicon accelerometer circuit.

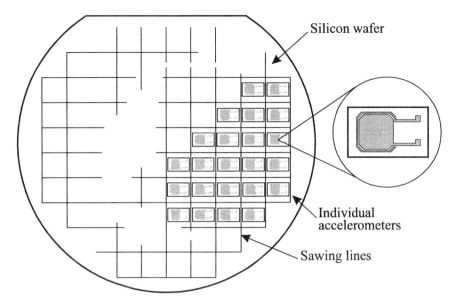

Silicon wafer

Individual
accelerometers

Sawing lines

Figure 4.8. Many accelerometers on a wafer.

glass plates to either side of the wafer, the pendulum motion is constrained so that
the hinge does not become overstressed and break.

It is possible to use the same electrodes for both sensing and forcing. The thin
metal film forming the combined capacitive pickoff and forcer is evaporated onto
the glass before bonding. Gas in the small gap between the mass and the glass
plates provides squeeze film damping. A servo loop between the pickoff and forcer
keeps the mass centered, as shown in Figure 4.7, and the forcer voltage is a
measure of acceleration. Potentials up to 50 V may be needed for $50g$
accelerometers, but there is virtually no power dissipated in the sensor. A Northrop
instrument with a range of $60g$ (at 15V) showed ±1mg day-to-day repeatability,
peak-to-peak, about the mean; 0.03% linearity; resolution better than $1\mu g$;
vibropendulosity less than $0.1\mu g/g^2$; and a bandwidth of 3kHz [20].

Another micromachined pendulous accelerometer, made by Draper Laboratories
and Rockwell, has a range of $100g$; bias stability of $0.55\mu g$ (1σ, 100 seconds); and
scale factor nonlinearity of 0.6% at $25g$ [22]. Also, Litton usessilicon
accelerometers with fiber-optic gyros (Chapter 12) in the LN-200 inertial reference
unit [23].

Micromachined sensors should cost an order less than the electromechanical
units they replace. Silicon is inexpensive and readily available for integrated
circuits. Mass-production equipment for the processes common to integrated
circuits, such as photolithography, dry etching, and metallizing, is sophisticated
and available with automated batch processing capabilities. IC wet etch equipment
can be readily modified for the micromachining operations, and the accelerometers
are cut from the wafer using an IC saw. But the main reason that micromachined
sensors are inexpensive is that each wafer can hold 80 or more accelerometers, as

shown in Figure 4.8. That means that they should cost less than $200 each (after packaging and testing) rather than $1000 or more paid for conventional electro-mechanical accelerometers. This will dramatically change the inertial sensor business, just as integrated circuits changed electronics.

References

1. IEEE STD 337-1972. Standard Specification Format Guide and Test Procedure for Linear, Single-Axis, Pendulous, Analog, Torque Balance Accelerometer.
2. IEEE STD 530-1978. Standard Specification Format Guide and Test Procedure for Linear, Single-Axis, Digital, Torque Balance Accelerometer.
3. Jacobs, E.D., "Accelerometer," U.S. Patent 3 702 073, 7 Nov. 1972.
4. Metzger, E.H., "Pendulous force-rebalance accelerometer," in Ragan, R.R. (Ed.) "Inertial Technology for the Future," IEEE Trans. on Aerospace and Electronic Systems AES-20, 4, 414–444, 1984.
5. Flamm, J., M. Hafen, B. Ryrko, B. Sinclair, "Development of a dry pendulum accelerometer at LITEF," DGON Symposium Gyro Technology, Stuttgart, 1982.
6. Shintani, Y., K. Sakuma, H. Yabe, H. Ito, K. Nishikawa, K. Kuramoto, T. Takahashi, "Development of a low cost high performance accelerometer," DGON Symposium Gyro Technology, Stuttgart, 1983.
7. Kariv, R., "Development of TM-74 TAMAM low cost high performance accelerometer," DGON Symposium Gyro Technology, Stuttgart, 1986.
8. Nicoli, J.A., "Perfect pendulous linear servo accelerometer model A834," DGON Symposium Gyro Technology, Stuttgart, 1986. The hyperbolic title should be taken as intent, rather than achievement; the paper describes investigations into error causes.
9. Smithson, T.G., "A review of the mechanical design and development of a high performance accelerometer," *Mechanical Technology of Inertial Devices*, Paper C49/87, Proc. Inst. Mech. Eng. (London), 1987.
10. Danielson, M.S., "Compensation of gain temperature coefficient in an optical pick-off for an accelerometer," U.S. Patent 4 598 586, 8 July 1986.
11. Petersen, K.E., "Silicon as a mechanical material," Proc. IEEE, 70, 5, pp. 420–457, May 1982.
12. Angell, J.B., S.C. Terry, P.W. Barth, "Silicon micromechanical devices," Scientific American, pp. 44–55, April 1983.
13. Satchell, D.W., "Silicon microengineering for accelerometers," *Mechanical Technology of Inertial Devices*, Paper C46/87, Proc. Inst. Mech. Eng. (London), 1987.
14. Roylance, L.M., J.B. Angell, "A batch fabricated silicon accelerometer," IEEE Trans. Electronics Devices, ED-26, pp. 1911–1917, 1979.
15. Youmans, A.P., "Solid state force transducer, support and method of making same," U.S. Patent 4 050 049, 20 Sept. 1977.

16. Block, B., "Solid state transducer and method of making same," U.S. Patent 4 071 838, 31 Jan 1978.
17. Stephens, M.L., Gray, P.R., "Temperature compensated piezoresistive transducer," U.S. Patent 4 166 269, 28 Aug. 1979.
18. Hansson, J.I., "Silicon accelerometer," U.S. Patent 4 553 436, 19 Nov. 1985.
19. Lawrence, A.W., "A microIMU using advanced inertial sensors," Proc. 14th Guidance Test Symposium, Holloman AFB, Oct. 1989.
20. Blanco, J., Geen, J., "Micromachined inertial sensor development at Northrop," ION, Proc. 49th Annual Mtg., Cambridge, MA, June 21–23, 1993.
21. Lefort, O., "A miniature, low cost, silicon micromachined servo accelerometer," DGON Symposium Gyro Technology, Stuttgart, 1988.
22. Barbour, N., et al., "Micromechanical silicon instrument and systems development at Draper Laboratories," AIAA Guidance, Navigation, and Control Conference, San Diego, CA, July 29–31, 1996.
23. Klass, Philip J., "Fiber-optic gyros now challenging laser gyros," Aviation Week & Space Technology, pp. 62–64, 1 July 1996.

5

Vibrating Beam Accelerometers

In this chapter we will investigate the design of vibratory accelerometers. Their underlying physical principle is that the transverse resonant frequency of a string or bar depends on the tensile stress, and the stress can be made a function of acceleration by fixing a proof mass to the string or bar. We will examine the design of two versions of vibrating beam accelerometers (VBAs) and describe their performance models. We will also consider a recent micromachined silicon version.

The Vibration Equation

If a mass hangs from a fine wire that is made to vibrate (Figure 5.1), its natural frequency, f_0, will vary as the acceleration in the direction of the string varies

$$f_0 = (1/2\pi)(ma/M)^{1/2}$$

where
    m = suspended mass
    a = acceleration
    M = string's mass per unit length

Although the Atlas missile (circa 1950) used an Arma vibrating string accelerometer, moving from this idea to a generally useful accelerometer has eluded designers until recently. Even the best-quality metals will relax a preset tension to some extent over time. In one accelerometer design, the proof mass was supported between two wire strings, and the sum of the two strings' resonant frequencies was kept constant by mechanically altering the tension in the string pair with a servo. The difference between the strings' frequencies was the measure of acceleration. In the early 1950s a single proof mass was suspended by three orthogonal pairs of strings in tension in an attempt to make a three-axis instrument, sketched in Figure 5.2. Electromagnets drove the strings to resonance using self-tuned oscillators; the difference between the frequencies of the two oscillators on each axis provided the x, y, and z output signals.

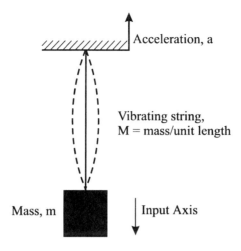

Figure 5.1. A vibrating string.

This accelerometer was unable to provide the in-run bias stability of a few μg needed for the 1 nm/h platform navigators that were then under development. The inadequate performance was due to the damping of the metal strings, to the

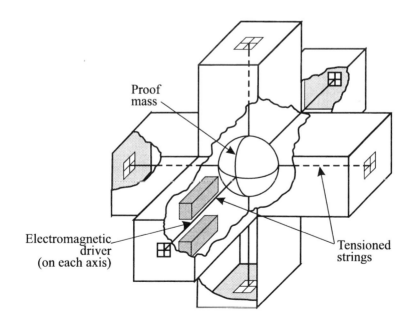

Figure 5.2. A three-axis vibrating string accelerometer.

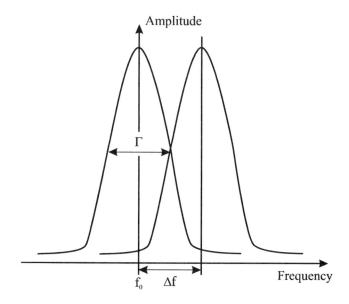

Figure 5.3. Resolution of two frequencies.

uncertainties in their attachments, and to the metal's creep (plastic flow), which lowered the resonant frequencies over time. As the two strings on either side of the proof mass on an axis were not identical, the drop in frequency was not the same for each string as the tension relaxed, leading to a bias change over time. Arma's instrument compensated for this change by modeling it.

The Resolution of a Vibrating Element Accelerometer

The resolution of a vibrating element accelerometer is its ability to measure small differences in frequency, which depends on the sharpness of the resonances. Sharpness depends on the damping of the oscillations and is described by the *quality factor*, Q. Equation (3.4) relates Q with the resonance linewidth and the resonant frequency, shown in Figure 5.3

$$Q = f_0/\Gamma$$

The resolution between two almost-equal frequencies with the same Q can be defined in terms of their Q. We can say that we can differentiate between them and that they are resolved when the frequency difference between them is equal to their linewidth Γ. (This is a kind of Rayleigh criterion). Then,

$$\Delta f = f_0/Q$$

It would be unusual for one of these accelerometers to have a sensitivity as high as 100 Hz/g; if it was, the theoretical resolution of this example would be only 0.1g. However, electronics techniques can be used to interpolate, and actual resolution can be better than the theoretical limit by one or two orders, depending on measuring time available and other conditions.

Despite the difficulty of achieving the resolution necessary for navigation systems, the search for the right way to make vibrating accelerometers continued, spurred by the advent of strapdown systems and laser gyros. They were sought for their three virtues:

1. They have an inherently digital output, unlike force rebalance types, which need to be electronically digitized.
2. They dissipate very little power, which does not increase with acceleration as electromagnetic forcer power does.
3. They reach their operating state in a fraction of a second, just as soon as the resonances are excited.

Interestingly, the higher the Q, the longer it takes for an external drive to excite the resonator oscillation to a steady state. In a high Q resonator there can be only a weak connection to the outside world. The excitation time constant τ is the reciprocal of the linewidth

$$\tau = 1/\Gamma = Q/f_0$$

Thus a resonator with Q = 30,000 operating at 30 kHz would take 1 s to reach its steady state if driven from the outside. For some applications (an air-to-air missile inertial reference unit, for example) this would be too long, so the resonator must be driven internally by a scheme that provides negligible damping losses.

In a resonator made from bars, the resonant frequency depends not only on the cross-sectional area and length but on the material's elastic modulus. One big advantage of bars over strings is that bars can operate in compression, whereas you cannot push on a string! Another advantage is that the bars do not need to be pre-tensioned as strings do, which removes the tendency to bias inducing creep.

It is difficult to make the resonant frequency independent of temperature because the elastic modulus of most materials changes with temperature. However, crystalline quartz resonators can be made from special crystal cuts that have negligible temperature sensitivity, so quartz has become the best material for "macroscopic" vibrating accelerometers (as distinct from micromachined silicon devices).

The Quartz Resonator

Successful vibrating accelerometers have been developed because single-crystal quartz has become readily available. It is piezoelectric, so it can easily be internally electrically driven to vibrate, and it is mechanically stable. Quartz crystal resonators had been used as electronic frequency standards since before World War II, but not in the quantities necessary to make them affordable for inertial instruments. The technology of quartz crystal resonators has advanced tremendously since microcircuits made the digital wristwatch inexpensive and commonplace. Quartz is now well understood, in vast production, and cheap. Using the same modern photolithographic techniques as those used to make silicon accelerometers (Chapter 4), watch crystal fabrication methods can easily be adapted for use in vibrating accelerometers.

Quartz resonators are stable over time and temperature, and quartz has little internal damping, which gives high-Q resonance peaks and high resolution. Quartz resonators for timekeeping can have Qs of 10^5 to 10^7; such crystals in an accelerometer with a resonant frequency about 10 kHz would have a theoretical resolution of 0.01 Hz, allowing mg or better resolution.

Two United States companies have developed quartz beam accelerometers, Astronautics Kearfott Corporation and Sundstrand (now AlliedSignal). Sundstrand's device is known as the "Accelerex." Generically we will call both instruments vibrating beam accelerometers (VBAs).

VBAs in General

VBAs are sensors in which a pendulum is constrained by a vibrating element. VBAs use two quartz resonators in push-pull, one being compressed by the acceleration and the other being tensioned. Each resonator is part of the feedback circuit of an electronic oscillator whose frequency depends on the stress in the crystal; the frequency of a tensioned resonator increases, while that of a compressed resonator falls. The acceleration is determined by differencing the two frequencies. To the extent that the two resonators have identical characteristics, differencing removes residual temperature sensitivity, aging, and even-order nonlinearities.

Because VBAs are open-loop instruments, they can have vibration rectification errors if the proof mass displacement is significant, which can happen in a pendulous design at the mechanical resonance of the proof mass on its hinge.

One might suspect that a device containing two independent oscillators could suffer from *lock-in*. This phenomenon (discussed as it affects ring laser gyros in Chapter 13) causes two lightly coupled oscillators, whose frequencies differ by a small amount, to lock to a single frequency. They stay locked up to a threshold when they suddenly unlock, and at larger difference frequencies each oscillates at its proper frequency. However, the VBA relies for its low sensing threshold on each resonator having an extremely high Q, meaning that each resonator couples extremely weakly to its surroundings. Hence the two oscillators couple extremely weakly to one another, giving a small lock-in band. Data from Sundstrand confirm that this is so; the lock-in band for an Accelerex is immeasurably small.

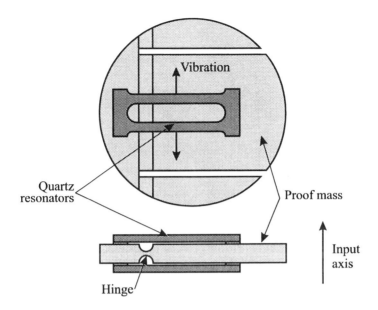

Figure 5.4. Schematic view of the Accelerex.

The Accelerex Design

The Sundstrand (AlliedSignal) Accelerex [1, 2] uses a double tuning fork resonator whose tension or compression depends on acceleration acting on a proof mass supported by a hinge. Sundstrand's illustrations [3] show two such resonators in push-pull, on either side of a single hinge, as sketched in Figure 5.4. The hinge provides a single degree of freedom for rotation about the output axis; it is made from metal, presumably for low cost. The proof mass is squeeze-film damped and its natural resonance is at 1450 Hz, where it has a Q of 3.4 (not to be confused with the *resonator* Q, which is 10^5 higher). Under external vibrations near 1450 Hz the vibropendulous error could become significant.

Each resonator is rectangular with the center cut away to leave two bars along the longer side, joined at the shorter ends by pads used for mounting, shown schematically in Figure 5.4. The resonator is 0.3 in. (7.6 mm) long; it can be thought of as a pair of tuning forks with their tines joined. Tuning forks are dynamically balanced, which means that the mounting end experiences no net forces because the tines vibrate in opposite directions. Each resonator is cemented at one end to the proof mass while the other end is anchored to the instrument frame; the stability of the cement joints determines the performance stability.

In another development, Sundstrand has blended one-piece fused quartz Q-Flex technology with push-pull Accelerex resonators to make a higher-accuracy accelerometer called "Superflex" [4]. It is not a pendulous design; the proof mass translates along the input axis (although it moves very little), substantially reducing

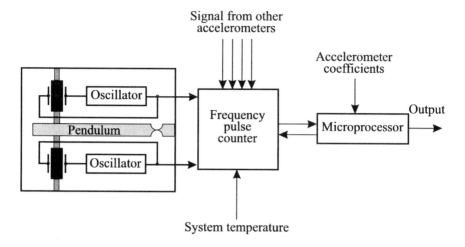

Figure 5.5. Accelerex system block diagram.

cross-coupling errors. The amorphous quartz flexure takes up a double bend (an inflection), keeping the proof mass parallel to its original plane, except for small rotations needed to compensate for thermal expansion differences between the resonators mounted on each side of the hinge.

Accelerex Signal Processing

The Accelerex has two built-in oscillator circuits, one for each resonator, and outputs their frequencies as buffered square waves. Figure 5.5 shows a typical data interface. Each resonator has a nominal center frequency of 35 kHz, which can change $\pm 10\%$ in response to acceleration, i.e., ± 3500 Hz for $\pm 100g$, or 35 Hz/g. In passing, note that if we guess that the resonator Q is 35,000, we can see that the unit will have a time constant of about 1 s, which means that it might be immune to short power interruptions if the output electronics maintains its phase reference.

With a scale factor of 35 Hz/g and the need to measure to 1 mg in at most 1 s, we can see that we must measure 0.035 Hz in 1 s or less. This is not a frequency measurement but a phase measurement, done by providing a 10-MHz system clock as a vernier to interpolate between difference frequency pulses (to measure phase). Thus scale factor accuracy will depend on the accuracy of the system clock. With this approach the resolution at 500-Hz sampling rate (2-ms samples) is 25 mg/sample, and as this improves linearly with time (assuming no lost counts) 1-s samples will give 50-µg resolution.

Sundstrand has various ways of compensating the Accelerex output. In one scheme the company modeled the output to the second order (using their terminology)

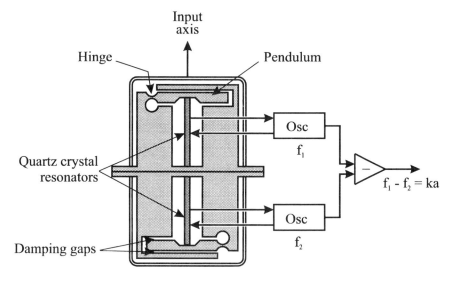

Figure 5.6. Schematic view of the Kearfott VBA.

$$a = a_0 + a_1 f_1^2 + a_2 f_2^2$$

where

$a_0 = \text{bias}$
$a_1, a_2 = \text{coefficients of second-order nonlinearity}$
$f_1, f_2 = \text{resonator frequencies}$

Terms a_0, a_1, and a_2 are then modeled to the third order in temperature

$$a_i = a_{i0} + a_{i1}T + a_{i2}T^2 + a_{i3}T^3, i = 0, 1, 2$$

Temperature varies relatively slowly, so compensation is done at much longer intervals than navigation computation; it can be done in a strapdown computer in the background, every second or so.

The Accelerex is a 100*g* full-scale instrument with 5 m*g* turn-on to turn-on bias stability. It is small, only about 0.6 in. in diameter (15 mm) by 0.6 in. long, including some of the output electronics. Sundstrand also offer an integrated circuit counter chip to make it easier to use.

The Kearfott Design

Kearfott uses a single vibrating beam in its resonator, according to Delaney et al. [5]. They say that Accelerex-like dual tine resonators have three problems:

1. They are more difficult to make because the tines have to be identical to have the same resonant frequency.
2. It is difficult to assemble the resonator so that both tines share the inertia force equally, because the resonator might be bowed or otherwise imperfect.
3. Having two tines sharing the inertia force means that the sensitivity is half that of a single beam or tine.

Kearfott's single-beam resonator provides the same dynamic balancing as a dual-tine resonator with a spring mass isolator. Albert's patent [6] describes one version of Kearfott's VBA; it is worth reading because it describes an elegant solution to a bias instability problem.

The VBA is shown schematically in Figure 5.6; the dual (push-pull) design has high common mode rejection of angular acceleration and anisoinertia errors. The design is quite straightforward, although assembly tolerances must be tight to align both half-accelerometers to the same reference axis.

Kearfott uses a cubic model to describe its VBA's resonator responses to acceleration. As usual, one resonator is in tension and the other is in compression. Using Kearfott's terminology, for Resonator 1 the output is

$$f_1 = K_{01} + K_{11}a + K_{21}a^2 + K_{31}a^3$$

For Resonator 2, the output is

$$f_2 = K_{02} - K_{12}a + K_{22}a^2 - K_{32}a^3$$

Differencing f_1 and f_2

$$
\begin{aligned}
f_1 - f_2 = & \ (K_{01} - K_{02}) && \text{Bias, nominally zero.} \\
& + (K_{11} + K_{12})a && \text{Scale factor.} \\
& + (K_{21} - K_{22})a^2 && \text{Nominally zero.} \\
& + (K_{31} + K_{32})a^3 && \text{Double, but each coefficient is small.}
\end{aligned}
$$

The tolerance on matching the two resonators determines the bias, and its stability will depend on their aging at the same rate. In practice, the difference frequency is at least 10 times more stable than each individual frequency. Bias and scale factor are claimed to be better than 10 μg and 10 ppm long term, and 1 μg and 10 ppm short term [7]. Long-term alignment stability of better than 1 arc-sec has been demonstrated. In contrast to Sundstrand's design, Kearfott's use of two separate pendula in opposition reduces the anisoinertia, vibropendulosity, and angular acceleration errors to very small levels.

Draper Laboratories has worked on a VBA called the Quartz Resonator Accelerometer (QRA). Its design is very similar to the one shown as Figure 7 in Albert's patent [6], in which the proof mass is suspended between two quartz resonators. The quartz is shaped using the micromachining techniques described

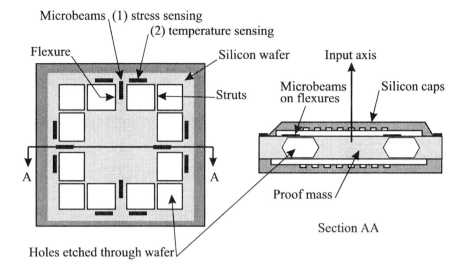

Figure 5.7. Honeywell's resonant microbeam accelerometer.

in Chapter 4 to make a very small instrument; otherwise it is the same as the two "discrete component" VBAs described earlier.

Silicon Micromachined VBAs

Silicon micromachining is being used to make resonant as well as pendulous accelerometers. Howe has described a three-axis unit in which four resonating beams support a square proof mass, the beams extending normal to the midpoints of the sides of the proof mass [8]. Because silicon is not piezoelectric like single-crystal quartz, the silicon beams are excited to resonance electrostatically. The third axis, normal to the plane of the proof mass, is servoed exactly like the micromachined silicon accelerometer (Chapter 4); this instrument is a mixture of open- and closed-loop techniques.

More recently, Honeywell has described a silicon VBA [9], shown schematically in Figure 5.7. A rectangular proof mass is supported by top and bottom cantilever flexures along four sides, and at each end it has supporting struts. In the center of the flexures on the top side there are *microbeams,* fabricated from polysilicon and electrostatically driven to resonate. Each resonator vibrates in a tiny vacuum enclosure with piezoresistors providing the feedback to the drive oscillators; they achieve Q-factors of 20,000. Operating each resonator in its own vacuum cell allows the proof mass to be squeeze-film damped. The microbeam resonators are placed at the outer edge of a flexure on one side and at the inner edge on the opposite side, such that one is under tension and one is under compression as the proof mass translates along the IA.

Under acceleration, the inertia force sets up a stress in the microbeam, changing its resonant frequency; the two microbeam frequencies move in the opposite sense. Then, each sensing beam has adjacent to it another pair of resonant beams, not connected to the proof mass, which are used as temperature sensors, providing compensation signals. The whole device is quite complicated; fabrication is described in [9]. Test data from prototypes give a scale factor of 700Hz/g on a 20g device. Honeywell suggests that milli-g stability and micro-g sensitivity are possible.

Comparison of Free and Constrained Accelerometers

The free servoed pendulum accelerometers (SPAs) are typified by the Q-Flex and the micromachined silicon accelerometer (MSA) (Chapter 4), whereas the constrained pendula are the VBAs and the QRA. They can be considered for two system performance classes, the 1-m/h aircraft and the 10-m/h aided navigators. First, let us make a general comparison of the two sensor types.

General Comparison of the SPA and VBA

The VBA output is inherently digital, whereas the SPA must provide A-to-D conversion, a potential source of error. But the analog SPA has potentially higher resolution, limited by electronics servo noise (usually low). The VBA electronics must avoid aliasing by using high sample rates, requiring a faster processor than the SPA.

The scale factor of the SPA varies with the aging of a permanent magnet and depends on there being a precise voltage reference. The VBA scale factor depends on the stability of the resonator materials—known from timekeeping to be sub-ppm over a year—and on accurate timing.

Because the VBA is open-loop, it has lower power consumption than an SPA at high accelerations, when the SPA would suffer thermal transients. But the servo stiffness of the SPA can better cage the pendulum under vibrations at its transverse natural frequency, when the VBA can suffer vibropendulous rectification.

The SPA pendulum is unsupported when unpowered, whereas the VBA pendulum is always constrained. This makes the VBA more rugged, tolerating higher levels of handling and storage shocks.

The SPA range is easily varied, and the servo characteristics are readily tailored to system needs, whereas the VBA is difficult to modify and tailor, requiring redesign for different full-scale ranges, for example.

Both instruments use the same fabrication technology, and both need to be assembled to the same cleanliness. Contaminants in the operating gaps, particularly in the damping (and pickoff) gaps cause stiction and poor resolution.

The VBA has the potential for lock-in at accelerations where the two resonators have the same frequency. This could cause scale factor and resolution problems not found in the SPA.

Comparison of Performance Ranges

Broadly, the 1-m/h aircraft and the 10-m/h aided navigators need accelerometers with the performance shown in Table 5.1.

The SPAs (Q-Flex and others) fill the 1-m/h system needs, and the VBAs are attempting to penetrate that market. The 10-m/h systems are be served by the VBAs, and perhaps the MSAs and QRA when they have completed development. One would expect VBAs, MSAs, and QRAs to be cheaper than the SPAs.

Table 5.1. Accelerometer Performance

	1-m/h aircraft IMU	10-m/h aided IMU
Range, g	30	50 (unmanned)
Bias stability		
Day-to-day, µg	60	N/A
In-run, µg	1	100
Scale factor stability, ppm	150	N/A
Nonlinearity, ppm	5	100
Asymmetry, ppm	5	100
Axis alignment, µr	80	500

Conclusion

Because of their digital output and relative insensitivity to angular motions, the VBAs should be excellent partners to optical gyros in modern tactical strapdown systems. The pendulous accelerometer is certainly not passé, though, because the MSA should be less expensive and smaller.

References

1. Norling, B.L., "Accelerometer with floating beam temperature compensation," U.S. Patent 4 718 275, 12 Jan. 1988.
2. Norling, B.L., C.J. Cornelius, "Accelerometer with isolator for common mode inputs," U.S. Patent 4 766 768, 30 Aug. 1988.
3. Norling, B.L., "Emergence of miniature quartz vibrating beam accelerometer technology for tactical navigation and flight control," Joint Services Data Exchange, Cambridge, MA, Oct. 1988.
4. Norling, B.L., "Superflex: A synergistic combination of vibrating beam and quartz flexure accelerometer technology," Navigation, J. Inst. Nav., 34, 4, pp. 337-353, Winter 1987–88.
5. Delaney, R., W. Albert, R. Weber, "Vibrating beam accelerometer," DGON Symposium Gyro Technology, Sttutgart, 1983.
6. Albert, W.C., "Monolithic resonator vibrating beam accelerometer," U.S. Patent 4 804 875, 14 Feb. 1989.
7. Albert, W.C., "Vibrating beam accelerometer," in Ragan, R.R. (Ed.) "Inertial technology for the future," IEEE Trans. on Aerospace and Electronic Systems, AES-20, 4, pp. 414–444, July 1984.
8. Howe, R.T., S-C Chang, "Resonant accelerometers," U.S. Patents 4 805 456 and 4 851 080, 21 Feb. 1989 and 25 July 1989. Patents differ only in claims.
9. Burns, D.W., R.D. Horning, W.R. Herb, J.D. Zook, H. Guckel, "Resonant microbeam accelerometers," Transducer '95, Eurosensors IX, 8th International Conference on Solid State Sensors and Actuators, Stockholm, Sweden, June 25–29, 1995.

6
The Principles of Mechanical Gyroscopes

In this chapter we present the base for describing, in subsequent chapters, different ways of making mechanical gyros. We will derive the Law of Gyroscopics and the expression for the Coriolis acceleration, the phenomena underlying spinning wheel and vibratory gyroscopes.

To study the earth's rotation, the French scientist Leon Foucault used a large pendulum (67 m long), which he built in the Panthéon in Paris. Its iron bob weighed 28 kg, and it swung with a period of about 15 s. As the earth rotated under the swinging pendulum, the plane of the swing appeared to rotate clockwise (because he was in the northern hemisphere) at vertical earth's rate because the pendulum's momentum was fixed in inertial space.

The pendulum motion was perturbed by draughts, so it soon becomes inaccurate as an inertial reference. So, building on the experiences of Bohnenberger (1810, Germany) and Johnson (1832, United States), Foucault devised a more compact and accurate instrument in 1852, a gimbaled wheel that, because of its angular momentum, would stay fixed in space while his laboratory rotated around it [1–4]. (Onlookers see the gyro rotating, because they are oblivious to their motion with the earth.) Foucault joined the Greek words that mean "to view" and "rotation," *gyros* and *skopein*, and coined the word *gyroscope* for his suspended wheel. While some writers reserve this word for rotation sensors with spinning wheels, in this book we use it for any instrument that signals or measures rotation. To begin, let us consider the inertial property arising from the angular momentum of a spinning wheel.

Angular Momentum

Angular momentum stabilizes. The familiar rifle is a gun with a spiral groove down the inside of its barrel, which spins the bullet to stabilize its path. A spinning top stays upright by virtue of its spin.

From Newton's Second Law of Motion we know that the angular momentum of a body will remain unchanged unless it is acted upon by a torque and that the rate of change of angular momentum is equal to the magnitude of the torque, T:

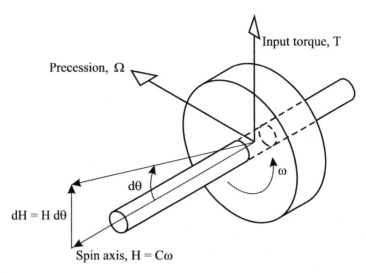

Figure 6.1. The law of gyroscopics.

$$T = dH/dt \tag{6.1}$$

where H = angular momentum = $I\omega$.

If the torque acts about the axis of rotation, its effect is to increase the angular velocity:

$$T = C \, d\omega/dt = C\alpha$$

where
   C = moment of inertia about the spin axis (polar moment)
   α = angular acceleration

In passing, note that the moment of inertia (which describes the *distribution* of the mass in a body) need not be constant, as in the case of the pirouetting ice skater (p. 93).

The Law of Gyroscopics

If a torque is orthogonal to an object's spin axis it cannot change the angular velocity vector magnitude, but it can change its direction, shown in Figure 6.1. The angular momentum H lies along the spin axis, and the torque T acts upwards, that is, it is a motion trying to turn the wheel in the direction that a right-handed screw would turn if it were moving along the same direction as T. The (small) angular momentum change dH is in the same direction as T, and its magnitude is

$dH = H \, d\theta$

where $d\theta$ = angle turned through. Combining this with (6.1) we get the Law of Gyroscopics:

$$T = dH/dt = H \, d\theta/dt = H\Omega \tag{6.2}$$

where Ω is the *precession rate*, the angular velocity of the wheel about the axis normal to the plane of the spin and input torque.

Conversely, if the wheel is turned about the precession axis, a torque appears about the orthogonal axis. To remember in which direction a gyro wheel moves, the rule is that when you try to turn a gyro wheel, it tries to turn so that its spin axis is aligned with your input torque.

Parasitic Torque Level

We are usually concerned with reducing gyroscope drift caused by parasitic torques, so let us visualize the size of a typical torque. A gyro for a 1-m/h navigator could have $H = 10^6$ dyn·cm·s (gm·cm²/s), and it needs to have random drift below 0.01 deg/h. Then

$\Omega = 0.01$ °/h $= 4.8 \times 10^{-8}$ rad/s

$T = 4.8 \times 10^{-2}$ dyn·cm

A dyne is roughly 1 mg weight; 0.05 mg is the weight of a piece of aluminum oven foil that is 1 mm square. Such a weight at a radius of 1 cm would cause this level of drift.

The Advantage of Angular Momentum

One could use the inertia of a stationary wheel to measure the rotation of a body attached to its spindle by measuring the angle between the spindle and the wheel. Once might try to economize in making a gyro by leaving out the drive motor and its power supply, merely supplying a pickoff or encoder to measure the angle between the shaft and the rotor. So now let us emphasize the benefit we get from spinning a gyro wheel. Assume that the friction in the bearings transmits a torque T to the wheel. In time t, the wheel's acceleration α leads to an angular velocity Ω_1

$\alpha = T/C \;\; \Rightarrow \;\; \Omega_1 = (T/C)t$

The same torque applied normal to a wheel spinning at angular velocity ω gives a precession angular velocity Ω_2

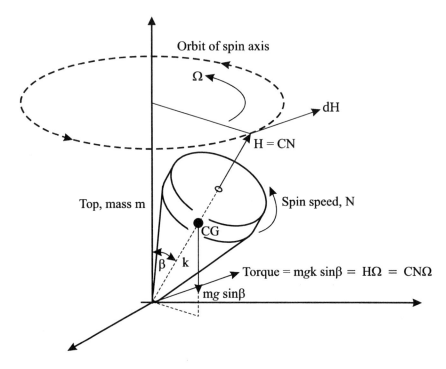

Figure 6.2. Precession of a spinning top.

$\Omega_2 = T/C\omega$ so that $\Omega_1/\Omega_2 = \omega t$

If T represents an unwanted parasitic torque, its effect is ωt times greater on the unspun wheel, and ω can easily be 2500 rad/s; angular momentum stabilizes far more than stationary inertia.

The Spinning Top—Nutation

A spun top stands upright, its spin axis making a slow coning motion in space about the vertical (it is *precessing*). Figure 6.2 shows such a top, with gravity acting to create a horizontal torque to topple the top. As we saw earlier, the spin axis will move to try and align itself with the torque, that is, it will move in the direction dH (parallel to the torque) in Figure 6.2. But as soon as it moves, the direction of the torque also moves because it is fixed in the top, so the precession direction changes, with the result that the top axis orbits in a circle. As the top slows, H decreases, and the top falls over when dH/dt can no longer balance the gravity torque.

For high spin rates, the precession rate is determined by the gravitational torque rather than by inertial torques. But if you were to tap a spinning top lightly with your finger, its spin axis would execute a rapid cycloidal motion called *nutation*,

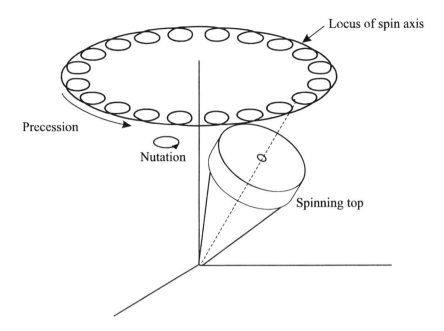

Figure 6.3. Precession and nutation.

as sketched in Figure 6.3. Nutation is less influenced by the gravity torque and is determined by the inertia forces acting on the spinning body.

Equations of Spinning Body Motion

To justify these assertions, let us assume a spinning body symmetrical about its spin axis, so that the transverse inertias are equal. If the body spins about the z-axis in a case with a fixed axes set x, y, and z, it has angular momentum $H = C\Omega_z$. We can write the torques about the x- and y-axes as the algebraic sum of the inertia torque and the gyroscopic torque (6.2)

$$T_x = A(d\Omega_x/dt) + H\Omega_y$$
$$T_y = A(d\Omega_y/dt) - H\Omega_x$$

where A = transverse moment of inertia.

This pair of differential equations has solutions that are the steady precession and an oscillation

$$\Omega_x = -T_y/H \qquad\qquad\qquad\qquad\qquad \text{Precession}$$

$$\Omega_y = T_x/H$$

$$\Omega_x = K_1 \sin \omega_n t + K_2 \cos \omega_n t \qquad\qquad \text{Nutation}$$

$$\Omega_y = -K_1 \cos \omega_n t + K_2 \sin \omega_n t$$

where

$$\omega_n = H/A = (C/A)\omega_z = \text{nutation frequency} \qquad\qquad (6.3)$$

K_1 and K_2 are constants found from the initial conditions. Note that for a sphere, $A = C$ and $\omega_n = $ spin speed. For a thin disk, $C = 2A$ and nutation is at twice the spin speed. Generally, the inertia ratio C/A for practical gyro rotors lies between the two extremes, and nutation occurs at about 1.5 times the spin speed.

The amplitude of nutation depends on the damping. In actual practice two-axis gyros (the only ones able to nutate) have small mechanical damping and need to be provided with nutation damping electronics.

Coriolis Acceleration

Gaspard de Coriolis, a French engineer and mathematician, postulated his acceleration in 1835, as a correction for the earth's rotation in ballistic trajectory calculations. The Coriolis acceleration acts on a body that is moving around a point with a fixed angular velocity and moving radially as well. We will describe it here because it drives the tuning fork gyro (Chapter 10), but it must also be compensated in navigation systems where the vehicle is moving over the rotating earth.

Figure 6.4 illustrates a mass at point P moving in the yz-plane so that it has a rotation rate ω about the x-axis and its radius (r) is varying. The mass velocity has components

$$dz/dt = -r \sin \theta \, (d\theta/dt) + \cos \theta \, (dr/dt)$$

$$dy/dt = r \cos \theta \, (d\theta/dt) + \sin \theta \, (dr/dt)$$

Differentiating each equation with respect to time and rearranging, we get expressions for the accelerations of P along OP and normal to OP. We assume that the angular and linear velocities are constant:

$$dr/dt = v \quad \text{and} \quad d^2r/dt^2 = 0$$

$$d\theta/dt = \omega \quad \text{and} \quad d^2\theta/dt^2 = 0$$

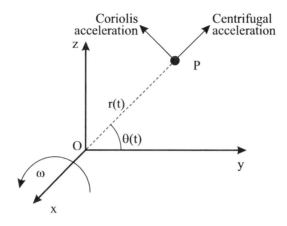

Figure 6.4. Coriolis acceleration.

and we have these accelerations:

Radial: $-r\omega^2$	Centrifugal acceleration	(6.4)
Normal: $2v\omega$	Coriolis acceleration	(6.5)

The centrifugal acceleration (6.4) causes a body to travel in a curved path because there is a force acting on it that is directed toward the path center of curvature. The sensation of centrifugal force that we experience in an automobile rounding a curve is the sense of our bodies attempting to obey Newton's First Law to continue in a straight line. The centripetal force (providing the turning) comes from the friction of the tires on the road, transferred to the car seat, and thence to our bodies. When distinguishing between centrifugal and centripetal forces we must consider which reference frame carries the observer. In the car, we carry it with us; when whirling a stone on a string and feeling the outward tension in the string, we are sensing the equal and opposite reaction.

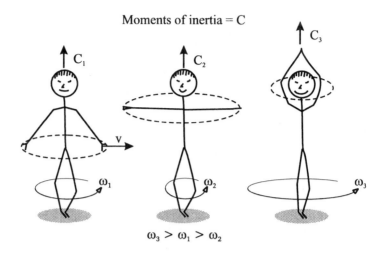

Moments of inertia = C

$\omega_3 > \omega_1 > \omega_2$

Figure 6.5. The ice skater.

The Ice Skater

It is well known that pirouetting ice skaters spin faster and faster as they move their arms from horizontal (a higher moment of inertia) to vertical (a lower moment of inertia), as shown in Figure 6.5. As there are practically no torques acting on the skaters and Newton's law requires that angular momentum must be conserved, when the skaters' moments of inertia fall, their spin speeds must rise. The Coriolis effect basically describes the acceleration acting when the moment of inertia is not constant, and it is this effect that causes the skaters' speed to change.

Gyroscopes with One and Two Degrees of Freedom

Mechanical gyros can be constructed so that the wheel is free to point to any direction in space by supporting the wheel in two gimbals, as shown in Figure 6.6. This is the type of construction that Foucault used in his gyro. As the wheel can rotate about two axes (in addition to its spin axis), we call this a *two-degree-of-freedom* gyro. It is also known as a two-axis gyro, a *displacement* gyro, or a *free* gyro. As we shall see in Chapter 8, this kind of gyro might not have physical gimbals; it may be made by electrostatically suspending a sphere, for example.

Around 1950 there was a school of gyro design on the U. S. east coast at Massachusetts Institute of Technology, led by Draper, that believed that for the highest accuracy, gyros should be restricted to one free axis, what we call a *single-degree-of-freedom* gyro, also shown in Figure 6.6. On the west coast, however, Autonetics pursued two-axis gyros with significant success; their system navigated

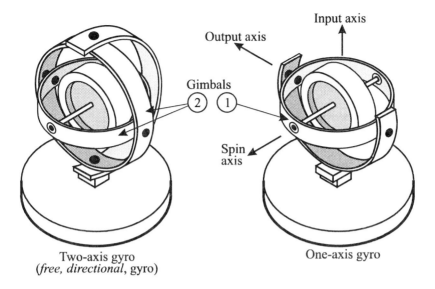

Figure 6.6. One- and two-axis gyros.

the *Nautilus* under the North Pole in 1958. Later, the less expensive dynamically tuned gyro (DTG), a two-axis gyro, became widely used on both coasts.

The choice between one-axis and two-axis mechanical gyros was not based on very powerful cost or performance criteria; each performed well and system costs did not seem to differ much. Even though wheel-type mechanical gyros are practically obsolete, they still may be preferred in some cases. In the next chapters we will describe the construction of both types of gyro, so that you can be better informed should you be faced with the task of deciding between them.

Conclusion

John Perry's book [5] on spinning tops is worth reading.

The essence of spinning wheel gyroscopes is in the Law of Gyroscopics, Equation (6.2)

$$T = H\Omega$$

It tells us that a gyro will precess or drift at a rate Ω under the influence of a torque T. The angular momentum H is therefore a primary design parameter for such a gyro; it should be as large as possible to keep drifts small. Gyro size strongly influences H because the wheel speed is practically limited to the 12,000 to 60,000 rpm range, whereas the polar inertia of a gyro wheel goes roughly as the fourth power of its diameter, making diameter the dominant parameter.

Equation (6.3) states that the free (two-axis) gyro will nutate with frequency

$$\omega_n = H/A = (C/A)\omega_z$$

The Coriolis acceleration, [Equation (6.5)]

$$a_c = 2v\omega$$

relates to bodies moving in an arc with varying radius.

References

1. Sorg, H.W., "From Serson to Draper—Two centuries of gyroscopic development," Navigation, J. Inst. Nav., 23, 4, pp. 313–324, Winter 1976–77.
2. Scarborough, J.B., *The Gyroscope—Theory and Applications*, Interscience Inc., New York, 1958.
3. Foucault, L., "Sur une nouvelle demonstration expérimentale mouvement de la terre fondée sur la fixité du plan de rotation," C. R. Acad. Sci., 35, p. 421, 1852.
4. Foucault, L., "Sur les phénoménes d'orientation des corps tournants entrainés par un axe fixe á la surface de la terre," C. R. Acad. Sci., 35, p. 424, 1852.
5. Perry, J., *Spinning Tops and Gyroscopic Motion*, Dover, New York, 1957. Library of Congress Cat. 57-3845. Reprint of lectures first published about 1900.

7
Single-Degree-of-Freedom Gyroscopes

In Chapter 6 we described the Law of Gyroscopics, so now we can see how instruments put this law to work and how they are designed. We begin with the single-axis, or single-degree-of-freedom gyro (SDFG). There are two types to consider, the rate gyro, an open-loop sensor, and the rate-integrating gyro, a closed-loop version. As usual, the closed-loop gyro has higher accuracy, whereas the open-loop gyro is less expensive. The most common SDFGs are filled with a fluid to provide damping and sometimes flotation; why and how this is done will be a subject of this chapter. The SDFG is the most common type in the field; more than a million have been put into service, and even though optical technology will make them obsolete, they will be in the inventory for many years.

An accelerometer can be made from a single-axis gyro in two ways, which we will also describe. One of them, the pendulous integrating gyro accelerometer (PIGA), is currently the most sensitive and accurate accelerometer available.

The Rate Gyro

A rate gyro provides a signal proportional to rate of rotation (angular velocity). The heart of the SDF rate gyro is a wheel running at high speed on low-noise ball bearings, usually driven by an electric motor. There are models used in guided weapons, where the wheel is spun up by a gas jet and allowed to coast, but because of the practical limits to the gas supply, these are useful only for short missions. Because the scale factor depends on the angular momentum, which is proportional to the wheel speed, gas-driven instruments can only be used where accuracy need be no better than a few percent.

The wheel is mounted in a frame or gimbal that is attached to the instrument case by one or two torsion bars; if one is used, the other end of the gimbal is connected to the case by a low-friction ball bearing. Figure 7.1 illustrates this construction; the torsion bar and bearing are on the instrument output axis (OA), the axis about which the gimbal turns in response to a rate about the input axis (IA). The wheel and motor assembly is sealed inside a housing filled with an inert gas such as helium, which allows the gyro to be filled with damping fluid. This housing forms the gimbal, to which are attached the torsion bar, an OA bearing,

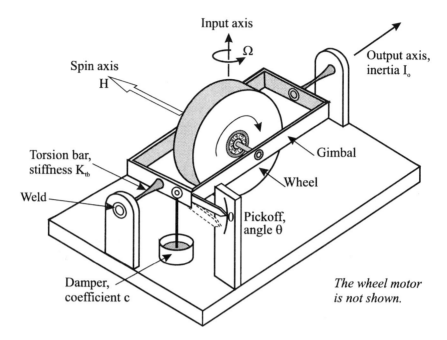

Figure 7.1. The single-axis rate gyro.

and perhaps the rotor of a damping compensator. A pickoff measures the gimbal angle and will provide an electrical output (usually AC) whose magnitude is proportional to the gimbal angle. A gimbal with its pickoff can be seen in Figure 7.2. Let us assume in the following that the pickoff has a sensitivity of K_{po} V/rad.

The Scale Factor

We can construct an expression for the gyro's scale factor using Equation (6.2), opposing the gyroscopic torque due to a steady rate Ω with a torsion bar of stiffness K_{tb} N·m/rad:

$$T = H\Omega = \theta K_{tb}$$

where θ = gimbal deflection. The pickoff provides an output signal, S, given by

$$S = \theta K_{po}$$

Therefore, the gyro scale factor [Equation (2.1)] is

$$K = S/\Omega = (K_{po}/K_{tb})H \tag{7.1}$$

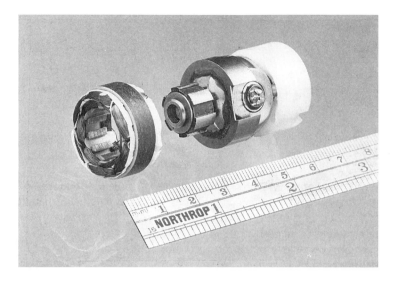

Figure 7.2. Pickoff and gimbal assembly. (Courtesy of Northrop Corporation.)

The angular momentum is the product of the rotor's polar inertia and its speed, and the moment of inertia is greatest for a given mass if the mass is concentrated in the rim (a large radius of gyration). The rotor mass should be as low as possible to minimize the vibration- and shock-induced loads in the wheel bearings, torsion bar, and OA bearing. Therefore the rotor is made of a dense material, and, to get its mass at the largest radius, it is arranged so that it surrounds the drive (spin) motor.

The Spin Motor

Mechanical gyros generally use a hysteresis motor, a type of motor also used in electric clocks because it is synchronous; i.e., its speed is locked to the frequency of the current that excites it. These motors have 2- or 3-phase windings so that they have a predetermined direction of rotation and a high efficiency [1]. In typical rate gyros this motor has a stator 0.5 in. (12 mm) diameter and 0.2 in. (5 mm) long wound with coils of fine wire. Typical motors generate enough torque to spin a wheel up to 24,000 rpm in a second; note that for a constant torque T and negligible drag torques

$$T = C\alpha = H/t$$

where
  C = rotor polar inertia
  α = angular acceleration
  t = run-up time; time to reach synchronous speed

For H = 10^4 dyn·cm·s and t = 1 s, T = 10^4 dyn·cm. Bearing drag and windage will increase this value.

The Ball Bearings

The rotor spins on ball bearings that are designed to be stiff enough so that the resonant frequency of the rotor and bearing assembly is greater than 1.5 kHz, outside the range of harmful environmental vibrations. The bearings are lubricated by small amounts of oil absorbed into a cage or retainer that rides around with the balls, keeping them apart. This oil must not solidify at the low temperatures specified for military equipment (as low as $-55°C$) or the wheel will not start. It must retain enough viscosity at high temperatures (80 to $100°C$) so that it properly lubricates the balls, preventing destructive metal-to-metal contact [2].

The condition of the bearings can be gauged by the wheel run-up and run-down times. Under a given motor voltage and at a given temperature the run-up time will depend on the drag on the wheel. Thus, if the oil is thickening or the bearing is running dry, the wheel will take longer to reach synchronous speed. When the power is shut off the wheel coasts to a stop; monitoring the run-down (coast) time will also indicate bearing torque. (If the run-up time is longer but the run-down time is normal or longer, one would suspect that the spin motor, rather than the bearings, has a problem. One might also check the spin motor supply electronics.)

Damping

Rate gyros are often liquid filled to provide damping for the correct dynamic response (page 102), and the damping fluid also serves to cushion the gimbal against shock and vibration. Damping is provided by viscous shear between the gimbal and housing parts; its magnitude depends on the fluid viscosity and the gap. However, fluid viscosity varies strongly with temperature, and rather than control gyro temperature, many gyros have mechanical damping compensators. By making the parts of materials with different coefficients of expansion the gap can be arranged to close down as the viscosity reduces with increasing temperature. The gap opens up as the instrument gets cold and the viscosity increases.

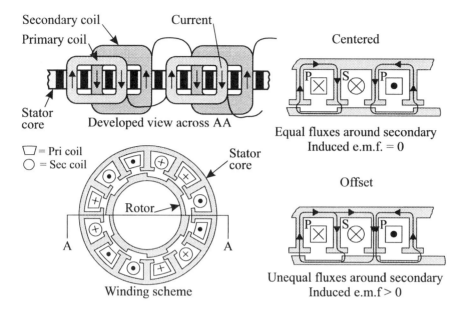

Figure 7.3. The microsyn principle.

The Pickoff

Often the pickoff is a *microsyn,* as shown in Figure 7.3, an electromagnetic type that has a toothed rotor on the gimbal and a stator attached to the case. The stator carries primary and secondary windings, the primary being excited by AC. The magnetic circuit is designed so that at null, magnetic fluxes generated by the primary cause no net flux linking to the secondary (Figure 7.3). When the rotor moves, the secondary flux is unbalanced and a voltage is generated whose magnitude is proportional to the angle and whose phase relative to the primary indicates the direction of rotation. Savet [1] describes the principles of microsyn design.

In an open-loop sensor the linearity of the pickoff determines the linearity of the instrument. Equation (7.1) shows that K_{po} should be as high as possible. The upper limit is set by the electrical noise threshold and by the torques from the increased magnetic fields in the pickoff, called *reaction torques*. Reaction torques modify the restraint about OA, changing the scale factor; they are unstable if the pickoff supply is not stabilized, and they can add magnetic hysteresis.

Finally, pickoffs must only measure the angle about the OA; they must not respond to radial displacement of the rotor in the stator.

The Torsion Bar

For high sensitivity, K_{tb} [in (7.1)] should be small; the weaker the torsion bar, the more it twists and the larger the output signal. But now the IA has rotated away from its original (reference) position, and the gyro will sense any rate component along the SA, a cross-coupling error. Therefore, stiff torsion bars are better because they give less cross-coupling. The stiffer the torsion bar, the higher the resonant frequency and bandwidth, and the better the gyro withstands shock and vibration. There is an upper limit to how stiff the bar should be, for the bar must be attached to the gyro housing, and the attachment must be much stiffer than the bar. The attachment must be very stable over time and temperature, because creep in the mounting between the torsion bar and the case frequently contributes to long-term null instability.

The torsion bar must be proportioned so that the stresses set up in it do not exceed its material's elastic limit; indeed, they must be low enough that hysteresis is negligible. High-strength alloys made for weighing scale springs, e.g., Elgiloy, make good torsion bars.

Flexleads

Flexible leads (*flexleads*) carry the spin motor power from the case to the gimbal. They must be rugged enough to withstand vibration and shock and must be able to carry the motor current without overheating, but they must be compliant enough not to add a significant torque to that of the torsion bar. Alloys like beryllium copper and coin silver, and gold, all in very thin ribbon form, can be used as flexleads. In a fluid-filled gyro (see later in this chapter) the flexleads must not react with the fluid or they will gradually deteriorate and break.

Rate Gyro Dynamics

We introduced the spring-mass dynamic system in Chapter 3 and showed that its motion was represented by a second-order differential equation. The rate gyro is also a second-order system; its torque balance equation about the OA is given by

$$I_o(d^2\theta/dt^2) + c(d\theta/dt) + K_{tb}\theta = H\Omega \qquad (7.2)$$

where
  θ = gimbal angle
  I_o = gimbal moment of inertia about the OA
  c = damping constant about the OA

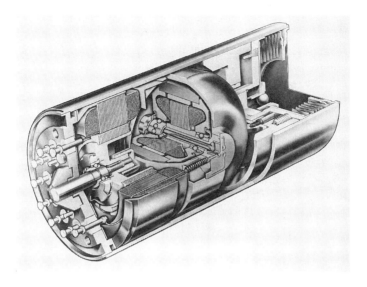

Figure 7.4. The Northrop GRG5 gyro. (Courtesy of Northrop Corporation.)

K_{tb} = torsion bar spring constant
H = wheel angular momentum
Ω = input rate

In (7.2) the gyroscopic torque $H\Omega$ is opposed by the inertia torque I_o times the angular acceleration about the OA, a damping torque c per unit angular velocity times the precession rate about the OA, and the torsion bar torque, $K\theta$. The damping ratio ζ and resonant frequency ω_n are given by Equation (3.3)

$$\zeta = \frac{c}{2(I_o K_{tb})^{\frac{1}{2}}}$$

$$\omega_n = (K_{tb}/I_o)^{\frac{1}{2}}$$

We saw in Chapter 3 that the instrument's bandwidth is limited by its resonant frequency, being defined by the 3-dB point on the amplitude plot or the $\pi/2$ phase lag value.

The IEEE standard [3] provides a complete specification of the SDF rate gyro and its error model. Edwards [4] describes some other error sources, and Simons [5] provides more information on their design.

A typical mass-produced rate gyro, the Northrop GRG5, is shown in Figure 7.4. (Northrop-designed instruments are now produced by AlliedSignal.) It is 1" in

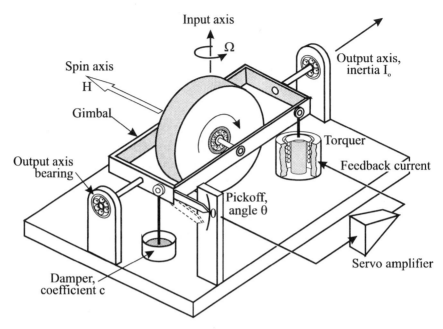

Figure 7.5. The rate-integrating gyro.

diameter and liquid filled; the electric spin motor can be seen in the gimbal, the damping compensator is at the back, and the torsion bar is at the front.

The Rate-Integrating Gyro

We can overcome the cross-axis sensitivity and resolution problems of the rate gyro by operating it closed-loop, in the same way as the Q-Flex and silicon accelerometers (Chapter 4). We remove the torsion bar and replace it with a torque motor, commonly called a *torquer*. The pickoff output now drives a servo amplifier, which is shown schematically in Figure 7.5, that supplies a current to the torquer. The torquer output torque is equal and opposite to the gyroscopic torque. The gyro output signaling the input rate is the torquer current (or a digital pulse stream, Chapter 3), not the pickoff angle, which in the steady state is null.

The Torquer

Torquers are often of a type named after d'Arsonval, who made a moving coil current meter by suspending a current-carrying coil between the poles of a magnet. When a current flows in the coil, it tries to align itself so that its plane lies normal to the magnet flux. By using a radial magnet flux, shown in Figure 7.6, the torque

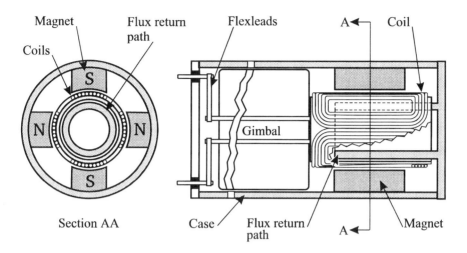

Figure 7.6. The d'Arsonval torquer.

may be constant over a wide angle. We make an SDFG torquer by attaching the magnet to the gyro case and the coils to the gimbal; it is bad practice to put the magnet on the gimbal, since stray fields will cause spurious torques. However, because we have put the torquer coil on the gimbal we must provide two more flexleads to carry the current to it.

Alternatively, we can use a microsyn as a torquer. Passing a constant current through one winding and a variable current through the other will generate a torque on the rotor. Such torquers can be used digitally, with binary or ternary pulse streams instead of variable currents. The microsyn torquer can provide an output of some tens of degrees a second, enough for a platform system but not enough for a strapdown system where rates easily reach 400 deg/s.

So we have replaced the torsion bar with a torquer; why does that mean that we must change the gyro's name, from rate to rate-integrating? The answer goes back to an original use of the SDFG in platform systems (Chapter 1). When an SDFG is used to stabilize a platform gimbal its pickoff output goes (via a servo) to the platform torquer, which then moves the platform to null the gyro. There is a torquer in the gyro, but it is used only indirectly, because by displacing the gyro gimbal and generating a pickoff signal we can turn the gimbals to a new angle. We might do this either during initial alignment or to maintain the gimbals aligned to the local vertical.

Because the gyro has no significant spring restraint about its OA, $K_{tb} = 0$ in Equation (7.2) and

$$I_o(d^2\theta/dt^2) + c(d\theta/dt) = H\Omega \tag{7.3}$$

The solution to this equation is

$$(\theta_o/\theta_i)(t) = (H/c)[1 - \exp(-(c/I_o)t]$$

The steady-state solution when t is very large is

$$\theta_o = (H/c)\theta_i$$

that is, the angle turned by the (gyro) gimbal is proportional to the input angle. The proportionality constant, H/c, is called the *gyro gain*. Because the output angle is proportional to the integral of the input angle, this gyro is called the *rate-integrating gyro*. Fernandez and Macomber [6] provide a rigorous analysis of the dynamics of the SDFG.

When a rate-integrating gyro is used in a strapdown system its pickoff is connected to its torquer through a high-gain electronic servo. Then it behaves like a rate gyro with a very high stiffness torsion bar.

The Output Axis Bearing

For the gyro to have a low threshold, the gimbal must respond to very small torques. Let us calculate the level of such a torque. A typical gyro wheel of about 1-in. (25-mm) diameter, spinning at 24,000 rpm, will have an angular momentum of about 10^5 dyn·cm·s (or g·cm^2/s). Let an input rate (Ω) be 1 deg/h, the same as 1 arc-sec/s, or 5×10^{-6} rad/s. The gyroscopic torque [Equation (6.2)] is

Torque = $H\Omega$ = 0.5 dyn·cm

Now 0.5 dyn·cm is the torque generated by a weight of 0.5 mg at a radius of 1 cm (because $g = 981$ cm/s^2, close enough to 1000 for this calculation). Half a milligram is the weight of a piece of aluminum oven foil about 3 mm (0.125 in.) square, not a very large amount; for the gimbal to respond to such a small torque it must be supported in very low-friction OA bearings. These OA bearings must also define their axis precisely (to prevent cross-coupling errors), which implies that they must have little free play.

Early gyro builders used compass or watch jewel bearings made of sapphire, with tungsten carbide pivots typically 0.016 in. (0.4 mm) in diameter. There were some gyros built where the jewel bearings were mounted in piezoelectric disks that, when excited with AC, caused the jewels to vibrate along the OA, reducing the friction level noticeably because dynamic friction is less than static friction. Nowadays, subminiature ball bearings are used, made from sapphire rings and tungsten carbide balls, and assembled to submicron tolerances. Other types of OA bearing have been used successfully [7].

In a strapdown system the OA bearings are loaded by gyroscopic forces and by the unfloated gimbal weight. When an SDFG is rotated about its OA, a gyroscopic torque appears about the IA, which has to be supported by the OA bearings:

Gyroscopic torque = $H\Omega_{OA}$

OA bearing torque = Fd

where
   Ω_{OA} = rate about the OA
   H = wheel angular momentum
   F = radial force on an OA bearing
   d = separation of the OA bearings

Therefore

$$F = H\Omega_{OA}/d \tag{7.4}$$

Assuming Ω_{OA} = 400 deg/s = 7 rad/s, H = 10^5 dyn·cm·s, and d = 3 cm, then F = 230 g weight. This load could easily be the gimbal's dry weight.

The friction torque in an OA bearing depends on the load on it, and here is where Draper took an idea from magnetic compasses and earlier German work in gyrocompasses [8] and turned it to advantage. He applied Archimedes' principle to the SDFG.

The Principle of Flotation

The gyro wheel was sealed in a gimbal of such a volume that it displaced its own weight of a fluid; the gimbal was *floated*. Draper's group put the OA bearings on the axis of the gimbal, now called the *float*, and by taking the load off the bearings they were able to make their friction torque very low and to provide very sensitive gyros.

Figure 7.7 shows a float in a tank of liquid. For true neutral buoyancy, not only must the float weight equal the displaced fluid weight but the float's center of gravity (CG) must coincide with its center of buoyancy (CB), located at the center of mass of the fluid displaced. If these centers do not coincide axially and radially the gyro will be sensitive to acceleration; we know that a buoy in the sea floats upright because its CG is below its CB, the buoy axes aligning with gravity. To avoid this tendency for a gyro float to align its OA with an applied acceleration—that is, to ensure that the OA bearing friction is zero at both ends—the float must be balanced end-for-end; it must not tend to turn OA-up if placed horizontally in a bath of fluid. One end-for-end balances the float by moving weight from the heavy end to the light end of the float until balance is achieved.

However, at the same time the float must be rotationally balanced, otherwise it will turn about the OA under acceleration along the IA and the SA, a *g*-sensitive error known as *mass unbalance*. Therefore, weights must be distributed around the OA to achieve balance, at the same time that the CB is maintained aligned to the CG. The float balancing procedure requires that simultaneously:

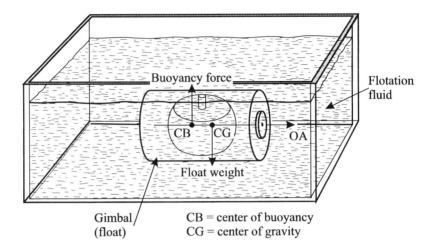

Figure 7.7. The flotation principle.

1. the float must be rotationally balanced out of the fluid so that its CG is along the OA;
2. the float mass must be adjusted to the exact value needed for neutral buoyancy at the chosen flotation temperature;
3. one must achieve end-for-end balance; and
4. one must achieve rotational balance in the fluid, so that the CB is also on the OA.

The technician does this by building the float a little light, noting how it floats, and using a computer program to tell where to place make-up weights around the outside of the float.

Damping

The flotation fluid provides damping by either of two mechanisms. In the first, viscous shear, the gap between the gimbal and the case is only 0.005 in. (0.15 mm) or so wide, and the fluid in the gap experiences cylindrical shear under rotation about the OA. Provided the flow is laminar (not turbulent), Newton's Law of Viscosity provides that the damping torque is proportional to the velocity, that is, c in Equation (7.3) is constant. The second mechanism, paddle damping, involves orifice flow; a case-fixed stator carries paddles with orifices and the gimbal carries paddles that mesh with the stator paddles. There is between 10 and 20° of angular freedom between them; when the gimbal moves it pumps fluid through the orifices.

Viscosity varies with temperature, the gyro gain depends on damping, and the strapdown gyro servo stability depends on there being enough damping. But tactical systems must be operating a couple of minutes from switch-on, leaving no time for gyros to be heated or temperature stabilized. They need internal compensation for viscosity change. Mechanical damping uses some kind of fluid valve whose restriction changes with temperature; it can be operated by the expansion compensating bellows' motion.

Alternatively, gyros can use a low-viscosity flotation fluid and augment the damping electrically by suitable currents in the torquer. Mechanical damping has the advantage that it does not expend power in the torquer as electrical damping does, but it is considerably more expensive and makes the gyro larger. This compensation does nothing for misflotation, of course.

The damping fluid adds a small inertia to the gimbal dry inertia, as some of the fluid mass is viscous-coupled to the gimbal. This must be allowed for in time-constant (I_O/c) measurements.

Flotation, then, provides the following:

1. output axis bearing support,
2. low threshold,
3. gyro damping about the OA;
4. squeeze film damping for vibrations and shocks, and
5. damping about the IA for transient rates about the OA.

However, it contributes no support for gyroscopic loading from steady OA rates.

Flotation Fluids

What fluid can we use to float the gimbal? For the gyro to be small we need a dense fluid, so one immediately thinks of mercury. But the problem with mercury is making the float *small* enough, because mercury is denser than iron. And mercury is electrically conducting, making it difficult to transfer power to the motor and torquer through flexleads. Of course, there were those intrepid engineers who tried to use mercury, but such a gyro never went into production.

There is a class of fluids with density about 2 g/ml, the fluorocarbons, which are electrically insulating and make successful flotation fluids [9]. As they must also provide damping, they must be available in a range of viscosities. Fluid density varies with temperature, and fluid expansion coefficients are higher than metals or ceramics, that is, the fluid density varies much more than that of the float, so the fluid buoyancy decreases as the temperature rises. Therefore a gyro is perfectly floated only at one temperature, and the higher fluid expansion makes it necessary to provide a bellows volume compensator. (The bellows can operate a damping compensating paddle damper, though.)

In the 1960s and 1970s it was considered normal to heat gyros to a set temperature, typically about 60 to 80°C (140 to 180°F); they were then always at flotation temperature and had a constant gain. The systems then in use in aircraft

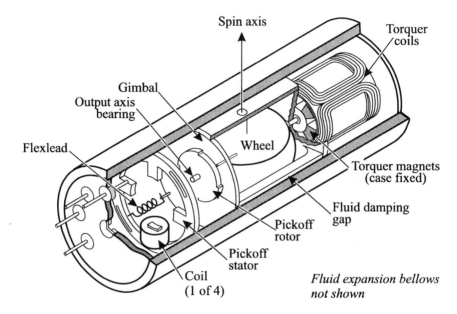

Figure 7.8. Single-axis floated gyro schematic.

and ships could afford to wait while the gyros warmed up and could supply the necessary power, which continues to be the case in strategic defense systems. Tactical strapdown systems are not heated, so SDFGs used in them are imperfectly floated.

The brominated fluorocarbon fluids can have densities up to 3 g/ml, but they can be chemically reactive; they have been known to eat flexleads away. They are expensive and have to be processed with care; therefore, they are avoided if at all possible. Silicone fluids are used for inexpensive gyros; they have the density of water (1 g/ml), and such gyros are sometimes not fully floated. The resulting performance is not as good as it could be if they were fully floated, but the price/performance balance is right for many tactical midcourse guidance systems, and for seeker head stabilizing gyros, among others. Silicone fluids have a smaller change in viscosity with temperature than fluorocarbon fluids, are chemically inert, and are cheaper.

Figure 7.8 shows a sketch of an SDF, rate-integrating floated gyro, and Figure 7.9 is a cutaway of Northrop's GIG6 SDFG [10]. Note the bellows-operated damping compensator.

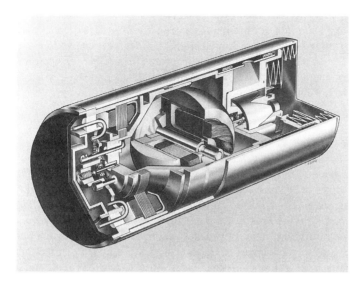

Figure 7.9. The Northrop GIG6 rate-integrating gyro. (Courtesy of Northrop Corporation.)

Structural Materials

Desirable properties for the structural material for SDFGs are mechanical stability, high thermal conductivity, low thermal expansion, and low density.

Mechanical stability is necessary so that the float CG stays coincident with its CB over time, otherwise the g-sensitive coefficients will not be stable. The tendency of materials to change shape over time is known as creep and is alleviated by symmetrical design, careful machining, and stress-relieving heat treatment.

We need high thermal conductivity because there is a heat source—the spin motor—in the middle of the float. Thermal resistances cause temperature gradients in the float, which, coupled with thermal expansion and any structural asymmetry, displace the CG from the CB, with the same results as creep.

The effects of temperature gradients can be minimized by choosing materials with low thermal expansion, although there are restrictions on the choice of materials for expansion coefficient, because we must match the float structure to the materials we must use, such as the spin motor magnetic iron.

We can benefit from materials with low density. For a given size of gyro and a given flotation fluid the float mass is defined, and for the highest gyro performance we want as much of the mass in wheel inertia as possible. Therefore we need a light gimbal structure, albeit stiff with good thermal conductance. Beryllium is an excellent material, expansion matched to steel (iron), with a Young's modulus almost twice that of steel, a thermal conductivity three times that of steel, and as light as magnesium. It has very low creep. It is expensive, though, and only justified in high-performance gyros. Magnesium is used in some gyros, but it has a low Young's modulus. Aluminum is more often used in tactical gyros despite its expansion being twice that of iron. Ceramics, though expensive to machine, have been used to make high-performance gyros; beryllia has extraordinarily high thermal conductivity, is strong, and expansion matched to steel, whereas alumina does not have such a high conductivity but is cheaper.

The Externally Pressurized Gas Bearing Suspension

Engineers have used other kinds of OA bearings in SDFGs. The German team working on the V2 (who came to the United States at the end of World War II in 1945) designed single-axis gyros and accelerometers in which the sensing elements were supported by externally pressurized gas bearings. They were used in the Apollo program's Saturn booster rockets in the 1960s but later went out of use in the United States. Mackenzie [11] reports that the Soviet Union used them as their main approach to inertial instrument design. (Output axis damping would be supplied electrically via a servo.)

Gyros have used bearings in which the flotation fluid was pumped around to make a hydraulic bearing (as in the crankshaft of a car engine); others have used taut wires tensioned in ingenious ways. One successfully deployed OA bearing is the magnetic suspension pioneered by Draper Labs and used in the gyros for the Polaris submarines and the MX missile guidance system.

A Magnetic Suspension

The magnetic suspension uses two microsyn-like wound stators each carrying four windings, shown in Figure 7.10. In order to provide axial as well as radial support, the stators and rotors have a conical gap. The eight coils are tuned by capacitors so that they are in series resonance at a chosen frequency, and the circuit has a high enough Q that the current changes markedly with frequency; this determines both the bearing stiffness and its operating range. Each circuit's natural frequency depends on its inductance, and the inductance depends on the gap between the stator and rotor. When the rotor is centered, the capacitors are chosen so that each circuit resonates at the same frequency. They are adjusted to work at the half-power point, obtained by decreasing the capacitance from the value at resonance. Then, if one leg's inductance falls (which it would do if the gap increases), its circuit moves toward resonance, the current in the coil increases, and the magnetic pull on the rotor increases. This is a negative feedback circuit and will reach an equilibrium position where the forces along each radial axis are balanced and the net axial forces from the four poles at each end are the same.

This is a passive suspension; it does not generate very much force, certainly not enough to oppose the OA rate loading in a strapdown SDFG. There have been servoed (active) suspensions made where an amplifier feeds current to the coils in

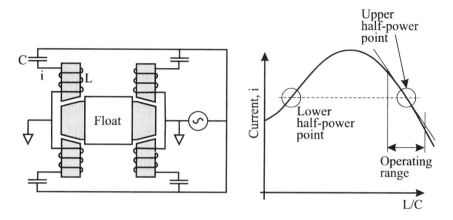

Figure 7.10. The passive magnetic suspension.

response to a radial/axial pickoff signal, and these are much stiffer and can provide enough force even for a strapdown gyro.

Self-Acting Gas Bearings

The ball bearing is inexpensive and rugged, and it consumes little power. But it is difficult to lubricate inside a floated gimbal and therefore the ball bearing is prone to wear out. If the spin bearing is overlubricated, the excess oil can move about inside the gimbal and change its state of balance, causing intolerable drift uncertainties; if underlubricated, the balls and track wear, and friction and noise increase to the failure point. Even when properly designed, fabricated, and installed, ball bearings generate a wide spectrum of vibration noise. What is more, the vibrations from the bearing at one end of the shaft can be at a slightly different frequency from those at the other end. These two sets of frequencies beat together, giving low difference-frequencies within the system bandwidth, noises that interfere with the navigation system.

The way around these difficulties is to use a self-acting gas bearing. In a gas bearing, the fixed and rotating parts are separated by a thin film of gas whose viscosity generates the pressure necessary to withstand a load. Because the viscosity of gases is so much lower than fluids, the clearances in a gas bearing must be much smaller than in a fluid bearing. There are a few different types of gas bearing [12]; a common one, the *spool* bearing, is made from a journal bearing and a pair of thrust bearings, as shown in Figure 7.11.

The journal bearing comprises a shaft that fits into a hole bored into the rotor with 1 to 3 µm clearance. The shaft may be from 3 to 20 mm in diameter, and its length is typically 2 to 3 times the diameter. When the journal is loaded, the

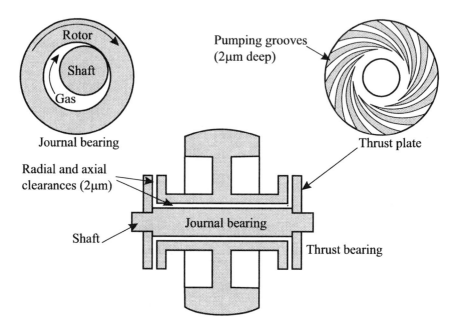

Figure 7.11. A spool-type self-acting gas bearing wheel.

clearance under the load gets smaller and the gas, dragged around by the viscous shear, is wedged into the reducing clearance. Even though the lubricating gas is compressible, the wedge action creates a pressure high enough to support the load. Such a bearing can be as stiff as a ball bearing, although it consumes more power than a ball bearing. Ausman [13] and Raimondi [14] describe journal bearing design.

The thrust bearing carries spirals some 2 μm deep etched into its surface (Figure 7.11). There is a bearing at each end of the shaft, spaced 1 to 3 μm from faces machined on the rotor; because of this small clearance the thrust plates and faces on the rotor must be flat to 0.1 μm and square to the journal to a 1 arc-sec. The thrust bearings are made with opposite-handed spirals so that when the rotor turns, viscous shear forces gas inward down the spirals and creates a pressure gradient that supports an axial load. Thus, the wheel can turn in only one direction. Gas bearings' theory and design are described in [15–17].

The only time that the bearing parts touch is during starting and stopping. As the surfaces are difficult to coat with a lubricant, the gaps being so small, the bearings are usually made from ceramic or other wear-resistant material. The bearings generate virtually no noise when running and can withstand operational vibration [18]. They have fewer parts than a ball bearing and are thus mechanically more stable; consequently, gyros using gas bearings are at least an order better in performance than ball bearing gyros.

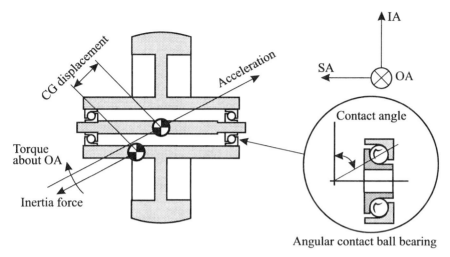

Figure 7.12. Ball bearing anisoelastic torque.

Gas bearings are expensive to make, but gyros using them last a long time, especially if the wheel can be left running. Gas bearing gyros are used in nuclear submarine guidance and in strategic missiles, where long operating life is necessary; they have been used in tactical systems (target designators), where extremely low-noise gyros have been needed for stabilizing a line of sight.

Anisoelasticity in the SDFG

In Chapter 2 we discussed anisoelasticity and saw how a torque could result when there was a component of the displacement of the center of mass of a structure at right angles to the inertia force [Equation (2.4)]. In the SDFG the rotor bearings (Figure 7.12) are the most compliant part of the gimbal, and they cause an anisoelastic torque about the OA. Therefore they must be made isoelastic; their net IA (radial) stiffness must equal their net SA (axial) stiffness.

Ball bearings are made with the balls running in grooves in an inner and an outer ring. These grooves or races are machined to such a size that when the bearing is assembled, the load vector forms an angle to the radius called the *contact angle*, illustrated in Figure 7.12 (*angular contact* bearings). The stiffness of the bearing depends on the preload of the assembly, that is, the force with which the bearings are pushed against one another when they are assembled into the wheel. The ratio of the radial to the axial stiffnesses depends on the contact angle. Therefore low gyro anisoelasticity depends on proper selection of the contact angle, after allowing for the small but not negligible compliances of the rest of the wheel and its support.

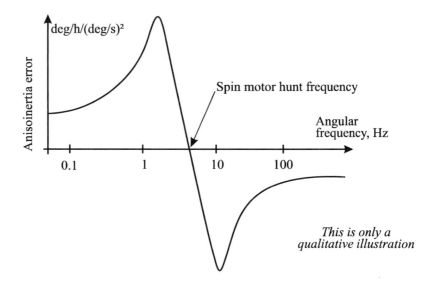

Figure 7.13. Anisoinertia in the SDFG.

In the gas bearing wheel, the stiffnesses of the journal and thrust bearings can be matched by choosing bearing parameters such as diameters, clearances, groove depths, type of gas, and gas pressure. But the journal itself is not isoelastic; when it is radially loaded, the center of rotation does not move along the axis of the load but moves off obliquely. This causes a *cross-anisoelastic* drift, which must be allowed for in gyro design.

Note also that the anisoelasticity goes as the square of the rotor mass, implying that light, high-speed rotors give better performance under high-acceleration.

Anisoinertia in the SDFG

Anisoinertia (Chapter 2) is only important in strapdown gyros and comes from the mismatch of the wheel moments of inertia around the IA and the SA. Consequently, the designer must distribute the rotor and gimbal masses to match these inertias.

Anisoinertia is frequency sensitive if the wheel is driven by a hysteresis motor, because hysteresis motors drive rotors around by dragging a magnetized ring on the rotor with a rotating magnetic field. This is an elastic coupling and is not very heavily damped; it has a natural frequency in the 4- to 40-Hz range. Gyro wheels will often *hunt* (their spin speed will oscillate) at this frequency.

Let us assume that the gyro is oscillated in the IA-SA plane, the frequency being swept slowly through the hunt frequency (Figure 7.13). At low frequencies the motor coupling connects the rotor inertia to the gimbal inertia about the SA, resulting in a value of anisoinertia from Equation (2.7), which we could call the "DC" value (because it would be the value appropriate for continuous rotation)

$$T_\Omega = \frac{1}{2}((I_{gs} + C) - (I_{gi} + A))\Omega^2 \sin 2\alpha$$

$$\Omega = \Omega_{max} \sin \omega t$$

where

I_{gs} = gimbal inertia about the SA
C = rotor polar inertia
I_{gi} = gimbal inertia about the IA (less rotor)
A = rotor transverse inertia

When the frequency ω of rate Ω is greater than the hunt frequency, the motor will not pass it, and the rotor inertia is disconnected from the gimbal inertia, giving a different value of anisoinertia, the *dynamic* anisoinertia, which will have a maximum value of

$$\Delta T_\Omega = \frac{1}{2}C\Omega^2$$

If this is a significant error source for a projected mission, one approach could be to design a special high-speed (say 96,000 rpm) wheel, which would have a low polar inertia for a given angular momentum and would give a smaller difference in DC-to-dynamic anisoinertia.

Vibration Rectification

The SDFG can rectify vibration components along the SA if they are exactly at the spin speed, N. The necessary condition is that the wheel rotor has a residual static unbalance, that is, its center of mass is radially displaced from the SA. Figure 7.14 shows a wheel with a radial unbalance, excited by a periodic motion along the SA with acceleration $a = a_o \sin (Nt + \phi)$. Imagine that the center of mass is in the plane of the figure above the SA, as illustrated, just when the input peaks ($a = a_o$) and its direction is to the left. Let the rotor mass be M, and the eccentricity of the center of mass be c. Then the rotor inertia will create an instantaneous torque clockwise about the OA

$$T_{OA} = Mea_o$$

Half a turn later the center of mass is in the plane of the figure and below the SA, and the acceleration has changed direction. Therefore the instantaneous torque is again clockwise

$$T_{OA} = M(-e)(-a_o)$$

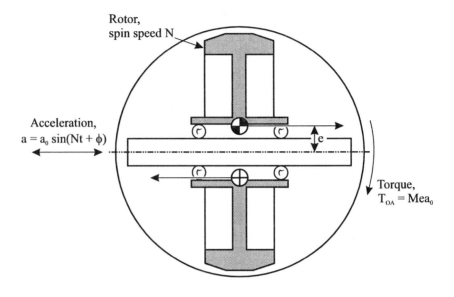

Figure 7.14. Axial 1N rectification error.

At the points where the unbalance is along the OA, the acceleration is zero so there is no torque. Therefore over a full cycle there is a net torque whose magnitude depends on the phase of the driving acceleration with respect to the position of the center of mass.

This error becomes important if two (or more) gyros are mounted on a platform or in a strapdown system and their rotors are not well balanced. As it is usual to drive all the gyros in a system from the same supply, their spin speeds will be identical, so one gyro creates a vibration at spin speed and the other rectifies it (and vice versa, of course). The gyro builder must be sure to balance each gyro wheel well, but the system designer must also be sure that the system does not have a structural resonance at the gyro spin speed, for then any unbalance accelerations can be amplified. Such a resonance can come from mounting the system on shock mounts, as has really happened, to the chagrin of many.

The SDFG Model Equation

The steady-state form of the IEEE model equation for the SDFG in a platform [19] is

S/K	Output signal/scale factor[1]
$= \Omega_i$	Input rate about the IA, in inertial space
$+ B$	Bias, non-*g*-sensitive drift rate
$+ D_i a_i$	IA mass unbalance times IA acceleration
$+ D_o a_o$	OA *g*-sensitivity times OA acceleration
$+ D_s a_s$	SA mass unbalance times SA acceleration
$+ D_{ii} a_i^2$	
$+ D_{ss} a_s^2$	
$+ D_{is} a_i a_s$	Acceleration-squared sensitive drift rates
$+ D_{io} a_i a_o$	
$+ D_{os} a_o a_s$	

The drift coefficients are read thus:
D_x = drift due to unit acceleration along x
D_{xy} = drift due to simultaneous unit accelerations along x and y

The accelerations used in the model equation are the components along the gyro axes of the vector sum of gravity and the motional acceleration (rate of change of velocity) of the gyro. On earth we define the apparent acceleration due to gravity as an upward motion of the gyro, so that its inertia force acts downward, in accordance with our experience.

For the strapdown gyro [20] we will need to consider terms arising from the gyro motion:

dS/K =	Motion-added error rate
$(I_s - I_i)\Omega_i \Omega_s / H$	
Anisoinertia error	
$+ \theta_o \Omega_s$	Cross-coupling error
$+ D_\Omega(\Omega_o)$	Rate error from rate about the OA
$+ I_o \alpha_o / H$	Error from angular acceleration about the OA

---

[1] The IEEE standard uses $-K_T$, the torquer scale factor, in this expression, defining it as the inverse of the gyro scale factor. We follow the conventions of Chapter 2 for consistency.

where

 H = wheel angular momentum

 I_s, I_i, I_o = moments of inertia about the SA, IA, and OA

 θ_o = angular displacement about the OA (pickoff output)

 $\Omega_{o,i,s}$ = rates about the OA, IA, and SA

 α_o = angular acceleration about the OA

We will describe how we measure these model coefficients in Chapter 15.

A Digression into Accelerometers

One can make an accelerometer by replacing the wheel in an SDFG with a mass whose CG is displaced along what was the wheel's SA. Under acceleration along the IA, the pendulosity generates an inertia torque about the OA, akin to the gyro's SA mass unbalance. By choosing the right mass one can achieve flotation, and by setting the pendulosity (the CG offset along the SA) one can set a scale factor that allows one to use the same servo loop as for the gyro version. In an inertial measurement unit with three accelerometers and three gyros, therefore, one can use six almost-identical instruments and six identical servos. In early versions of the AMRAAM air-to-air missile guidance system the six sensors used a single multiplexed servo, an arrangement that reduced cost considerably, because the gyros and accelerometers used most of the same parts and were assembled on a single production line. It must be said, though, that if one did not have an SDFG to start with, this would be an expensive way to build an accelerometer; it was only the parts' commonality that made it a viable IMU in its time.

The Pendulous-Integrating Gyro Accelerometer

In contrast to the previous gyro-based accelerometer, the PIGA is an accelerometer that uses gyroscopic action to convert an acceleration-induced inertia force into a torque that precesses a gyro [21, 22]. The acceleration is then proportional to the precession rate, and the angle that the gyro turns through is the time integral of the acceleration (proportional to the average velocity). The German V2, mentioned in Chapter 1, used an open-loop version of the PIGA [23].

 Figure 7.15 illustrates the PIGA; the gyro wheel (which might have gas bearings) is mounted in a floated gimbal, which is made so that its center of mass is offset along the SA, making it pendulous. The float carries a pickoff on its OA and has OA bearings in the usual way. The gimbal fits across the diameter of a housing mounted to the inertial platform through bearings along the wheel IA, and

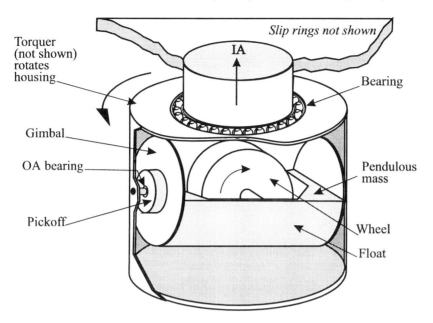

Figure 7.15. The pendulous integrating gyro accelerometer.

there is a torquer that can rotate the gimbal about the IA. Power passes to the gyro through slip rings.

When there is an acceleration along the IA, the gimbal pendulosity causes a torque about the float OA. The pickoff senses a small motion and its output drives a servo that turns the gimbal about the IA, generating a gyroscopic torque about the OA, which cancels the torque due to acceleration. Under a constant acceleration along the IA the gimbal steadily rotates about IA, and the *inertial* rotation rate is a measure of the acceleration.

PIGAs are very precise; they have a low threshold provided the gyro part has low stiction OA bearings (perhaps a magnetic suspension). They have excellent linearity and stability because their scale factor is determined only by the pendulosity and angular momentum of the gyro wheel; their ultimate accuracy depends on timing (rather than a voltage reference as in servoed sensors). Their dynamic range goes from 10^{-7} to 50 g or more. However, because they rely on gyroscopic torque to counteract acceleration, PIGAs cannot be used in strapdown systems where there can be rotations about the IA from vehicle motion. They are quite expensive, and their cost can only be justified in 0.01-m/h or better systems.

Conclusion

The SDFG is the best understood gyro in production, and this chapter has described only the skeleton of its theory and manufacture. There are quite a few different options for configuring the SDFG. For example, we mentioned (page 116) that hysteresis spin motors hunt around their synchronous speed, an effect that modulates the angular momentum. The *permanent magnet motor* (PMM), a class of brushless DC motor, is servo-controlled to a precise spin speed and does not hunt. If it is necessary to have a very quiet gyro, then the PMM and gas bearing combination is superior; it is much quieter than an optical sensor, and in some applications, such as stabilizing a line of sight (e.g., to point a laser beam) to nanoradian accuracy, it can be the only viable instrument. The PMM is also more efficient than a hysteresis motor, dissipating less heat into the gimbal and maintaining constant temperature gradients, but one drawback is that it needs complicated electronics, and is therefore more expensive than a hysteresis motor.

Whereas the optical gyros offer the advantages of longer life, no mass unbalance, and other benefits that we will list in later chapters, the standard SDFG in quantity-production costs less than a ring laser gyro (RLG), although the interferometric fiber-optic gyro (IFOG) may soon undercut its price.

References

1. Savet, P.H. (Ed.), *Gyroscopes: Theory and Design*, McGraw-Hill, New York, Chapter 11, 1961.
2. Holmes, J., "Rotor ball bearings for precision gyroscopes," *Symposium on Gyros*, Proc. Inst. Mech. Eng. (London), 1964–65, 179, 3E.
3. IEEE STD 292-1969. Specification Format for Single-Degree-of-Freedom Spring Restrained Rate Gyros.
4. Edwards, C.S., "The dynamic response of a rate gyro with imperfect rotor bearings and a viscoelastic damping fluid," in Willems, P.Y. (Ed.), *Gyrodynamics*, Springer-Verlag, New York, 1974.
5. Simons, W.R., "Miniature rate gyroscopes," in *Symposium on Gyros*, Proc. Inst. Mech. Eng. (London), 1964–65, 179, 3E.
6. Fernandez, M., G.R. Macomber, *Inertial Guidance Engineering*, Prentice Hall International, London, 1962. Good treatment of the SDF gyros. Appendices treat vibropendulosity and anioselasticity.
7. Trayner, B.T., "A survey of low torque gyroscope gimbal suspensions," *Symposium on Gyros*, Proc. Inst. Mech. Eng. (London), 1964–65, 179, 3E.
8. Sorg, H.W., "From Serson to Draper—Two centuries of gyroscopic development," Navigation, J. Inst. Nav., 23, 4, pp. 313–324, Winter 1976–77.
9. Coldwell, T., S. Scregg, O.H. Wyatt, "Consideration of precision gyro design using symmetrical rotors, including a discussion of flotation fluids," *Symposium on Gyros*, Proc. Inst. Mech. Eng. (London), 1964–65, 179, 3E.
10. Koning, M.G., "High volume single degree of freedom gyroscopes for strapped-down use," DGON Symposium Gyro Technology, Stuttgart, 1977.
11. Mackenzie, D., *Inventing Accuracy*, The MIT Press, Cambridge, MA, 1993.

12. Lawrence, A.W., "Comparisons between the theoretical performances of different designs of gas-bearing wheels," Proc. Conf. on Gyro Spin Axis Bearings, MIT, December 12–14, 1966.

13. Ausman, J.S., "An improved analytical solution for self-acting, gas lubricated journal bearings of finite length," Trans. ASME, J. Basic Eng., Series D, 83, 2, 1961.

14. Raimondi, A.A., "A numerical solution to the gas lubricated full journal bearing of finite length," ASLE Trans., 4, pp. 131–155, 1961.

15. Muijderman, E.A., *Spiral Groove Bearings*, Springer-Verlag, New York, 1966.

16. Grassam, N.S., J.W. Powell, *Gas Lubricated Bearings*, Butterworths, London, 1964.

17. Forse, R.W., A.G. Patterson, "Aerodynamic gas spin axle bearings for gyros," *Symposium on Gyros*, Proc. Inst. Mech. Eng. (London), 1964–65, 179, 3E.

18. Lawrence, A.W., "A gyroscope for satellite damping," *Symposium on Gyros*, Proc. Inst. Mech. Eng. (London), 1964–65, 179, 3E.

19. IEEE STD 517-1974. Standard Specification Format Guide and Test Procedure for Single Degree of Freedom Rate-Integrating Gyros.

20. IEEE STD 529-1980. Supplement for Strapdown Applications to IEEE Standard Specification Format Guide and Test Procedure for Single Degree of Freedom Rate-Integrating Gyros.

21. Wrigley, W., W.M. Hollister, W.G. Denhard, *Gyroscopic Theory, Design, and Instrumentation*, MIT Press, Cambridge, Mass., Chapters 13 and 14, 1969.

22. Ahn, Byong-Ho, "Floated inertial instruments," in Ragan, R.R. (Ed.) "Inertial technology for the future," IEEE Trans. on Aerospace and Electronic Systems AES-20, 4, pp. 414–444, July 1984.

23. Pearson, E.B., "An introduction to the gyroscope, an historic instrument," *Symposium on Gyros*, Proc. Inst. Mech. Eng. (London), 1964–65, 179, 3E.

8

Two-Degree-of-Freedom Gyroscopes

In the 1950s, when inertial navigation began, there were two schools of thought as to how to make gyros. To generalize somewhat, the East Coast school, led by Draper (at MIT), opted for single-degree-of-freedom floated gyros, which constrain the precession of a gyro wheel to a single rotational axis (the output axis). The West Coast school, however, successfully made two-degree-of-freedom gyros (2DFGs). Some companies (Litton, and East Coast Arma) made floated 2DFGs, while Autonetics made gimbal-less gyros using a ball-and-socket made from a spherical self-acting gas bearing inside the wheel. Others (Autonetics and neither coast Honeywell) made the ball-and-socket the entire gyro, supporting a sphere either electrostatically or magnetically, the latter type using superconducting materials to eliminate power dissipation and heating. Two-degree-of-freedom gyros are sometimes called two axis-gyros, or *free gyros*.

The Two-Degree-of-Freedom (Free) Gyro

In the ideal free gyro, the axis of a spinning wheel (or sphere) steadfastly points to a direction in space, undisturbed by restraints or torques. Foucault tried to make a free gyro by suspending his gyro's outer gimbal from a twistless silk thread and sitting the inner gimbal inside it on knife edges. He could not move his gyro about the room, but in a navigation system the gyros must be able to move in any direction, through any angle, without affecting the system's inertial reference stability. It is one thing to do this in a laboratory tabletop instrument, and another to do it in a mass-producible, small, inexpensive instrument that can withstand the environments of normal use.

As inertial measurement units (IMUs) usually need to provide rotation data in all three axes, they need three SDFGs or two 2DFGs. Generally, two 2DFGs give a smaller IMU of a given performance, although two 2DFGs provide four sensing axes, one more than is needed. This redundancy can often be turned to advantage in a strapdown system by configuring the four gyro input axes (IAs) at equal angles in space rather than along the familiar orthogonal axes set. In that way each IA senses components of rate from two vehicle axes, so that if one sensor axis fails, we still have information about all three system axes, providing extra reliability for

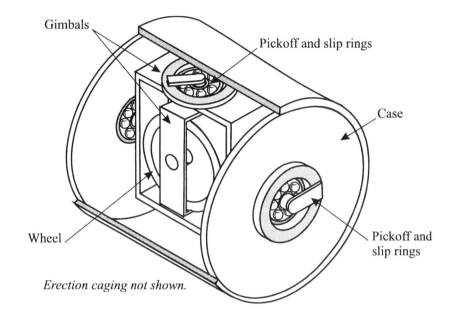

Gimbals

Pickoff and slip rings

Case

Wheel

Pickoff and
slip rings

Erection caging not shown.

Figure 8.1. A two-axis gyroscope.

long-lived systems. For this to be useful, though, the failure must affect only one axis, not the whole gyro.

In some applications, such as the gimballed radar or infrared seekers of tactical missiles, there is no need for roll rate measurement; only pitch and yaw must be stabilized. While it was common to use two single-axis rate gyros in such seekers, a single two-axis gyro is cheaper and smaller and more likely to be used nowadays.

Because it is the most common type of free gyro, we will devote the whole of Chapter 9 to the dynamically tuned gyro; in this chapter we shall consider some other types of 2DFG. We shall examine a simple external gimbal gyro first. Next, we shall describe the designs closest to realizing our free-gyro ideal, the electrically supported spherical rotor gyro, and its close relative, the cryogenic gyro. Finally, we shall mention the gas bearing free rotor gyro.

The External Gimbal Type

In some older missiles and in general aircraft instruments, one can find a gimballed two-axis gyro like the one sketched in Figure 8.1. The electrically driven wheel, mounted in gimbals, transmits gimbal angles through potentiometer pickoffs. Slip rings transfer power to the wheel and pickoffs, or, if limited angular freedom can be tolerated, flexible cabling might be used. Such instruments have a very low angular momentum for their size, because the wheel diameter must be small enough to fit inside the gimbals. Consequently, this kind of gyro does not have the performance needed for inertial navigation.

The expression for the nutation frequency [Equation (6.3)] of a free gyro must include the gimbal inertias and is given by

$$\omega_n = \frac{H}{(I_1 I_2)^{1/2}} = \frac{CN}{(I_1 I_2)^{1/2}} \tag{8.1}$$

where
   H = wheel angular momentum
   I_1, I_2 = gimbal inertias about axes orthogonal to H
   C = wheel polar moment of inertia
   N = wheel spin speed

The large inertias of the gimbals lead to a low nutation frequency and give this kind of gyro poor dynamic performance.

Two-Axis Floated Gyros

Both Litton Corporation and American Bosch Arma made successful 2DF floated gyros; Arma's was used in a ship's gyrocompass [1,2]. Litton (and others) sold thousands of 1-m/h aircraft navigators with 2DFGs. One can imagine the 2DFG as an SDFG with a spherical float, with output axis (OA) bearings connecting it to an intermediate gimbal ring rather than to the case. The gimbal ring lies concentric with the wheel and is made of magnesium or beryllium so that it is neutrally buoyant in the flotation fluid. The gimbal ring is attached to the case by another set of "OA" bearings, so that the spin axis, the inner gimbal axis, and the outer connection to the case are orthogonal as sketched in Figure 8.2.

Each degree of freedom needs its own pickoff and torquer; however, both components can be engineered with only a few degrees of angular displacement. Because it is impractical to provide high-power torquers in this kind of design, it is limited to platform use. Both electromagnetic and capacitive pickoffs have been used, as have moving-coil (d'Arsonval) permanent magnet torquers.

As in the SDFG, flexleads carry the current to the spin motor and to the torquers. Second-order servos typically connect the pickoffs to the inertial platform torquers, the gyro torquers are used to command the platform to align it or to maintain it in local level coordinates.

Also as in the SDFG, the use of a flotation fluid (to remove gimbal bearing loads) means that expansion bellows are needed. The free gyro does not integrate rate; the fluid viscosity is low (1 cP or so) and provides negligible damping to rotation, although it does provide some squeeze-film support to radial shocks. Consequently the 2DFG does not need the small-clearance damping gaps necessary

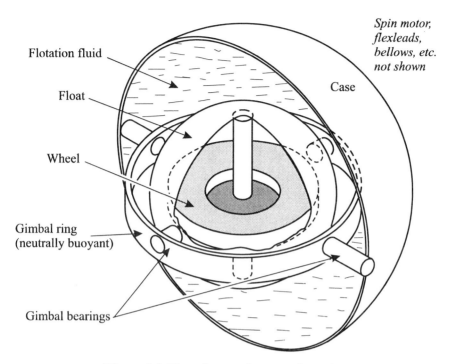

Figure 8.2. Floated two-axis gyro schematic.

in the SDFG, which are a potential source of gimbal stiction if the fluid is contam
inated. Nor does it need damping compensator mechanisms, as does the unheated
SDFG.

The spin bearings must be isoelastic, just as in the SDFG, but in a platform there
is no concern for anisoinertia because the gyros null the vehicle rotations. For the
same reason, there are no angular acceleration error torques.

While SDF rate-integrating floated gyros have been preferred over the 2DF type
in the most precise navigators (in ICBMs and nuclear submarines), neither type
would be used in a new 0.1–1-m/h navigation system design; optical gyros would
almost certainly be selected for reasons that we shall see in later chapters. Even in
tactical strapdown systems, the entrenched, mature, SDFG is being replaced by
fiber-optic gyros that promise the lowest cost of ownership.

Spherical Free Rotor Gyros

There is an inherent elegance in using a sphere as a gyro, for a freely supported
spinning sphere will retain its SA alignment in inertial space so that vehicle rotation
can be measured with respect to the SA. (As one cannot measure vehicle rotation
about the spin axis, two gyros with orthogonal spin axes are actually necessary for
defining the attitude completely.) Gyros have been made from spheres supported

electrostatically and magnetically; both Honeywell and Rockwell Autonetics have made versions of the electrically supported gyro (ESG). Research has continued at Stanford University, with the goal of making an ESG to be placed in an earth satellite to test general relativity. This gyro must have extremely low drift, no greater than 2×10^{-11} deg/h; one design had a superconducting pickoff [3].

In the 1960s, General Electric worked on a magnetically supported gyro using a cryogenic superconducting sphere of niobium, operating in liquid helium, the only superconductor available at that time. It was considered plausible to make an entire cryogenic IMU, with superconducting accelerometers and a gravity-gradiometer as well as gyros, but because of the difficulty of operating a system in a cryostat (the power demands for refrigerant would be exorbitant), no cryogenic system has yet made it out of the development labs. Simon [4] gives some interesting background on superconducting suspension.

With the discovery of high-temperature superconductors in 1987 we might expect a revival of the superconducting gyro, and it will be interesting to see what ingenious inventors do with these new materials, and whether they can compete in the market against the optical gyros now in development. However, they have not yet materialized as of 1998.

The Electrically Suspended Gyro

In 1954 Arnold Nordsieck invented the electrically suspended gyro [5], shown in Figure 8.3. ESGs have been used in very high-precision navigators; Rockwell's MICRON (*MICRO Navigator*) was designed for high-performance tactical aircraft, whereas Honeywell's GEANS (*Gimballed ESG Aircraft Navigation System*) is used in strategic bombers. ESGs are even used to monitor the performance of SDFGs in nuclear submarine's navigators [6–8].

As engineered for production, the ESG uses a beryllium sphere, made to 0.1-µm tolerances, supported inside a spherical ceramic cavity (not shown) by the electric fields from electrodes placed along the three principal axes. Servo loops control the electrode voltages, varying the electric field to keep the sphere centered. The cavity is evacuated, raising the electrical breakdown threshold between the electrodes and the ball, but there is still a limit to the voltage that can be applied. This limits the acceleration that the ESG can stand to about $20g$. Evacuating the spherical cavity also prevents air drag on the rotor, which would couple a disturbing torque from any case angular motion.

Coils generating a rotating magnetic field induce eddy currents in the rotor sphere and behave like an induction motor, making it spin. When the sphere has spun up to typically 150,000 rpm, the motor current is turned off and the rotor coasts; any residual drag is compensated for by phased voltages applied to the electrodes so that the spin speed stays constant. Currents injected in specially placed coils damp any nutation induced during spin-up; they create out-of-phase

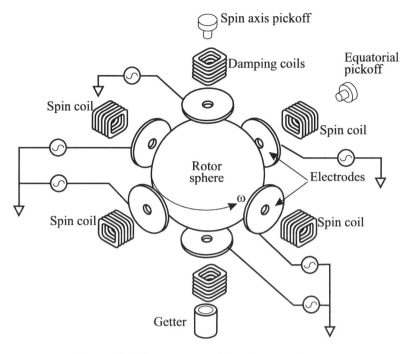

Figure 8.3. Components of the electrostatic gyro.

induced currents in the rotor. Note that the nutation frequency of a sphere, deduced from Equation (6.3) by setting $C = A$, is the spin speed.

Honeywell and Rockwell's designs differ in detail. Honeywell uses a hollow rotor 1.5 in. (38 mm) in diameter with an internal inertia ring to give it a well-defined principal moment of inertia, for the rotor must have a well-defined principal inertia axis so that its SA is stably located in the rotor (C is not exactly equal to A). Optical pickoffs measure the attitude of the outer case relative to the inertially stable rotor sphere. Rockwell uses a solid spherical rotor and a novel technique called *mass unbalance modulation* (MUM) for the pickoff. As sketched in Figure 8.4, Rockwell makes the material for its rotor by extruding beryllium, and in the billet Rockwell embeds tantalum wires (9 times the density of beryllium) off-center. This creates a preferred principal moment of inertia, and the mass center is radially displaced by 25 microinches (0.6 µm) from the ball's geometric center. This makes the ball center orbit in the electrostatic suspension, and the attitude of the ball is determined by combining the vectors of the unbalance signals from the capacitance-sensing (support) electrodes. Rockwell's gyro has a 0.4-in. (1-cm) ball running in a 300-microinch (7.5-µm) plate-to-rotor gap with a suspension drive of 150 V.

The ESG sphere must be round at operating speed, so to allow for the centrifugal strain dilation it must be fabricated slightly smaller around its equator than about its poles. The combination of rotor nonsphericity and electric field forces

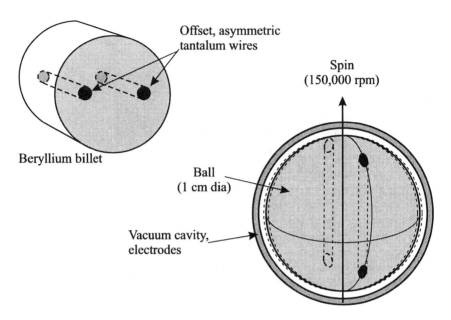

Figure 8.4. Mass-unbalance modulation.

determines the gyro bias drift, so this strain adjustment must be very accurate.

Free gyros have mass unbalance drifts just as SDFGs do. If the center of suspension does not coincide with the center of mass, the gyro will respond to gravitational or vehicle accelerations and will drift away from its original alignment. The center of suspension is defined by the electrodes; if they are not precisely formed they will induce torques causing the rotor to precess, which may differ for different case attitudes. ESG output is compensated for these effects with elaborate computer models, made possible because the drift coefficients are very stable; the ESG is capable of drift performance well below 0.0001 deg/h.

The precision of fabrication and the high quality needed to maintain the high vacuum make this a very expensive gyro, and its intolerance of high accelerations (its low *g*-capacity) limits it to navigation in large, smooth-riding vehicles like ships, submarines, and strategic bombers.

The Gas Bearing Free Rotor Gyro

A gyro can be based on a spherical bearing like a ball and socket joint, and the spherical bearing is usually a self-acting gas-lubricated bearing, as shown in Figure 8.4. Here the rotor has a high polar moment of inertia because of its flywheel shape, and it is spun by an induction motor. Unlike the ESG, this gyro has only a limited angular freedom, so it must be used in a platform. It is difficult to torque this design of gyro, and although the techniques we will see later for

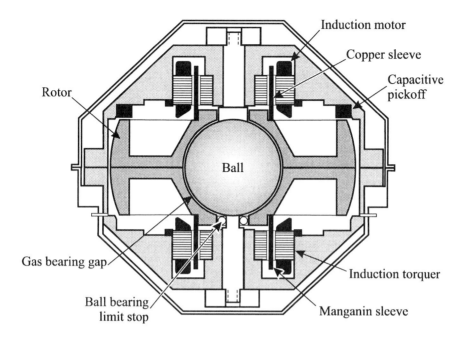

Figure 8.5. Gas bearing free rotor gyro.

torquing the dynamically tuned gyro could have been used, the design illustrated (based on Rockwell's G6 gyro) used an induction torquer. The G6 gyro has a capacitive pickoff. If the platform should slew too fast, the spinning rotor could hit the shaft, with dire consequences, so the shaft carries a ball bearing as a safety stop [9].

The G6 gyro is roughly spherical, 5.3 in. (135 mm) in diameter, and the gas bearing ball is 1.625 in. (42 mm) in diameter. The gas bearing clearance is 150 microinches (almost 4 μm), and it uses hydrogen gas at 20 psia as the lubricant. The wheel spins at 156 rev/s, 9360 rpm—relatively slowly. It can withstand a linear acceleration of 34g; this kind of gyro has a low shock tolerance. Despite this, almost 10,000 G6 and similar free rotor gas bearing gyros have been used in high-performance aircraft navigators, and they have given reliable service, with a mean-time-to-failure of more than a million hours.

The cryogenic gyro, the ESG, and the gas bearing gyro fail catastrophically if they are overloaded, because then the stationary and rotating parts rub at very high speed, generating high spot temperatures that fuse the parts together. Despite this lack of ruggedness, they have all been successfully used in platforms with better than 0.1 m/h performance.

References

1. Savet, P.H. (Ed.), *Gyroscopes: Theory and Design*, McGraw-Hill, New York, 1961.
2. Barnett, D., "The Arma-Brown two-axis floated gyro," *Symposium on Gyros*, Proc. Inst. Mech. Eng. (London), 1964–65, 179, 3E.
3. Everitt, C.W.F., "A superconducting gyroscope to test Einstein's General Theory of Relativity," in *Laser Inertial Rotation Sensors*, SPIE Vol. 157, pp. 175–188, 1978.
4. Simon, I., "Forces acting on superconductors in magnetic fields," J. Appl. Phys., 24, 1, Jan. 1953.
5. Nordsieck, A., "Principles of the electric vacuum gyroscope," in R.E. Roberson and J.S. Farrior (Eds.) *Guidance and Control*, Academic Press, New York, 1962.
6. Pondrom, W.L., "Electrostatically suspended gyroscope," and Hadfield, M.J., "Hollow rotor ESG technology," in Ragan, R.R. (Ed.) "Inertial technology for the future," IEEE Trans. on Aerospace and Electronic Systems AES-20, 4, pp. 414–444, 1984.
7. Huxley, A.S., J.D. Nuttall, D.C. Witt, "The electrostatically suspended gyroscope—A review of mechanical design aspects," in *Mechanical Technology of Inertial Devices*, Proc. Inst. Mech. Eng. (London), 1987-2. Paper C61/87 (27 refs).
8. DeNezza, E.J., R.R. Warzynski, R.L. Ringo, "The evolution of ESG technology," DGON Symposium Gyro Technology, 1972.
9. Pondrom, W.L., "Gas film supported free-rotor gyroscope," in Ragan, R.R. (Ed.) "Inertial technology for the future," IEEE Trans. on Aerospace and Electronic Systems AES-20, 4, pp. 414–444, 1984.

9
The Dynamically Tuned Gyroscope

In the 1940s engineers in Scotland designed a gyro that used a spinning flywheel on a universal (Hooke's) joint (Figure 9.1). The gyro was surprisingly unstable, and Arnold and Maunder at the University of Edinburgh showed that the dynamic inertia effects of the gimbal in the universal joint were responsible.

Twenty years later, Kearfott designed a gyro (the "Gyroflex") using a rotor on a universal joint with flexure pivots instead of bearings. Permanent magnets attached to the gyro case provided an attractive force on the rotor with a negative stiffness, the force decreasing as the distance between the magnets and the rotor increased. The magnetic negative rate spring canceled the flexure positive rate, freeing the rotor from torque, a necessary condition for a free gyro. However, the magnetic anti-spring suffered from erratic torques due, for example, to hysteresis, so that the gyro performance was inadequate for the market at that time.

Then, in 1963, Howe of American Bosch Arma received a patent for a gyro that used the dynamic inertia effects described by Arnold and Maunder to oppose the flexures in a Gyroflex-like gyro (instead of the magnets). The flexure spring stiffness is independent of spin rate but the dynamic inertia spring stiffness is not, so at a particular speed (the *tuned speed*) the two cancel. The resulting sensor is a dynamically tuned gyro (DTG), sometimes called a tuned rotor gyro or a dry tuned gyro.

The DTG Tuning Effect

Figure 9.2 shows the DTG's rotating parts; they comprise a rotor, flexures, a gimbal, and a shaft driven by a motor. The key parameters are the torsional spring stiffnesses of the flexures, the moments of inertia of the gimbal, and the spin speed. The gimbal is not necessarily a ring, but whatever its shape, it has polar inertia C_g and transverse inertias A_g and B_g.

Assume that the rotor moments of inertia A, B, and C are much larger than those of the gimbal, as they usually are. As the rotor assembly spins, driven by a motor on the shaft, the gimbal will rotate without twisting the flexures only if the rotor spin axis is aligned with the shaft. But if the rotor is offset (spinning in a fixed attitude), its spin axis is at a fixed angle to the shaft and the gimbal must flutter because it is connected to both the rotor and shaft by flexures that can twist but not

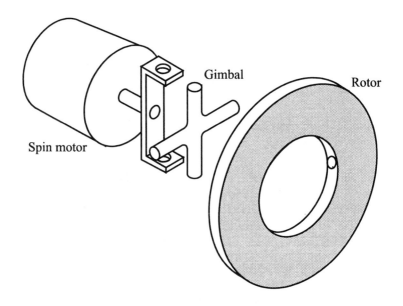

Figure 9.1. A universal (Hooke's) joint gyroscope.

bend. Therefore, at one position, the gimbal plane is forced to be normal to the shaft, while a quarter turn later it is forced to lie in the plane of the rotor (as shown in Figure 9.2).

The fluttering of the gimbal does two things. It winds up the flexures and produces a torque from their positive spring rate. Secondly, it puts angular rates on the gimbal which, because of its angular momentum, generate gyroscopic precession torques. These act like a dynamic negative rate spring about the flexure axes, opposing the flexure wind-up torque. The tuned speed is the speed at which the flexure torque equals the dynamic torque and the rotor is unrestrained.

The Tuning Equations

The equations for the tuning conditions were derived by Howe and Savet [1], Craig [2], and others [3]. There are five conditions to be met if a DTG is to be a perfect free gyro, in which the rotor is totally free of disturbance torques from small case motions. We will derive the expression for the tuned speed and quote the other four tuning conditions.

Figure 9.3 shows the parameters used to derive the expression for the tuned speed. Let us assume for now that we have frictionless pivots in the universal joint; they transfer torques only in bending. The pivots are also stiff, so that the gimbal only displaces about the inner pivot axis. Coordinate set (X,Y,Z) is fixed to the case, and set (x,y,z) rotates with the gimbal; the rotor rotates at angular speed N, so that the gimbal rotation angle at time t is $\theta = Nt$ (Figure 9.3). The rotor attitude,

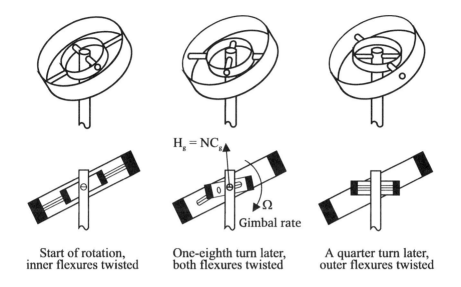

Start of rotation,   One-eighth turn later,   A quarter turn later,
inner flexures twisted   both flexures twisted   outer flexures twisted

Figure 9.2. Dynamically tuned gyro operation.

given by small angles α_x, α_y (in X,Y,Z), is constant over one revolution. The gimbal deflection, angle β, is found by transforming the rotor attitude from the fixed to the rotating axes

$$\beta = \alpha_x \cos \theta + \alpha_y \sin \theta$$

Transforming the gimbal angle back to case-fixed axes

$$\gamma_X = \beta \cos \theta = \alpha_x \cos^2 \theta + \alpha_y \sin \theta \cos \theta \qquad (9.1)$$

$$\gamma_Y = \beta \sin \theta = \alpha_x \cos \theta \sin \theta + \alpha_y \sin^2 \theta$$

To accelerate the gimbal we must supply an inertia torque and a Coriolis torque [Equation (6.5)] on each case-fixed axis

$$T_X = A_g(d^2\gamma_X/dt^2) + C_gN(d\gamma_Y/dt) \qquad (9.2)$$

$$T_Y = A_g(d^2\gamma_Y/dt^2) - C_gN(d\gamma_X/dt)$$

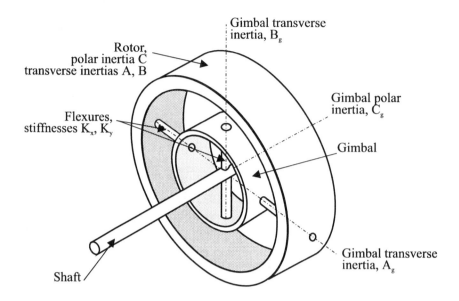

Figure 9.3. Dynamically tuned gyro parameters.

Because the outer pivot cannot transmit a torque from the gimbal to the rotor, only the torque on the inner flexure is transmitted. So, in rotating coordinates, the rotor torque is

$$T = T_X \cos \theta + T_Y \sin \theta$$

Transforming the rotor torque to fixed axes (X,Y,Z):

$$G_X = T \cos \theta = T_X \cos^2 \theta + T_Y \sin \theta \cos \theta$$
$$G_Y = T \sin \theta = T_X \sin \theta \cos \theta + T_Y \sin^2 \theta$$

Substituting from (9.1) and (9.2),

$$G_X = \tfrac{1}{2}A_g(d^2\alpha_x/dt^2) + A_gN(d\alpha_y/dt) - (A_g - \tfrac{1}{2}C_g)N^2\alpha_x \tag{9.3}$$
$$G_Y = \tfrac{1}{2}A_g(d^2\alpha_y/dt^2) + A_gN(d\alpha_x/dt) - (A_g - \tfrac{1}{2}C_g)N^2\alpha_y$$

plus terms in $\sin 2\theta$ and $\cos 2\theta$, which add to the rotor's own inertia and Coriolis terms (but we assumed the rotor was not precessing).

Now the third term in Equations (9.3) represents a negative spring, of stiffness $-(A_g - \tfrac{1}{2}C_g)N^2$, created dynamically by the gimbal flutter. If the pivots are replaced by spring flexures (whose stiffness is independent of spin speed), we can cancel their stiffness with the dynamic negative spring at the tuned speed. Figure 9.4

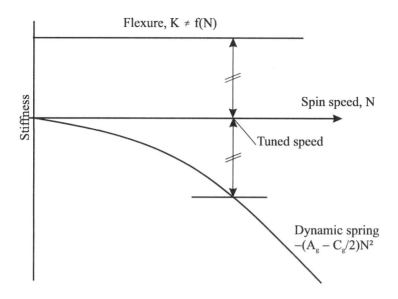

Figure 9.4. Tuning with a negative dynamic spring.

shows the parabolic form of the dynamic spring against spin speed and illustrates the tuned speed. If we allow for an asymmetric gimbal with transverse inertia A_g and B_g, and for different flexure stiffnesses K_x and K_y, the tuning condition requires that

$$(A_g + B_g - C_g) = K_x + K_y$$

The first of the five free-gyro operating conditions is the tuning condition

$$N^2 = \frac{K_x + K_y}{A_g + B_g - C_g} \tag{9.4}$$

Notice that this equation is analogous to that of the resonant frequency of a mass vibrating on a spring; the tuned speed is the square root of a stiffness term divided by a mass term.

Second, the condition for the rotor to be free in inertial space requires that there be no torques acting on the rotor from damping. This is an important condition, and it is discussed later.

Third, K_y should be zero. If it is not, and in practice it never is because designers choose K_x and K_y for other reasons (like anisoelasticity, discussed later), then the gimbal flutter is coupled to the rotor. Because the rotor inertias are so much greater than the gimbal inertias (see the figure of merit, later), the rotor motion is small. Since the gyro is usually operating at null, where there should not be any gimbal flutter unless the pickoff null is displaced from the gimbal null, this condition is not important, and we only mention it for the sake of completeness.

And the fourth condition requires that, for ideal tuning, it is necessary that

$$A_g + B_g - C_g = 0$$

which is satisfied if the gimbal is a laminar disk. But this condition cannot be met in practice, because it implies an infinite tuned speed.

Finally, the fifth condition is that the rotor, too, should be a thin disk

$$A + B - C = 0$$

where
    A, B = rotor transverse inertias
    C = rotor polar (spin) inertia

Because the rotor cannot easily be a thin disk and still carry a torquer, the penalty for not meeting this condition is that the rotor nutates.

DTG Nutation

The DTG has two nutation frequencies, one given by NC/A [Equation (8.4) for a symmetric rotor where $A = B$], and another given by $N(2A - C)/A$. In practice only the first (higher) nutation frequency is important. DTGs commonly have C/A values around 1.7, so the nutation frequency is about 1.7 times the spin speed. Any disturbance causes a DTG to nutate, because it is a stable normal mode; the servo electronics must counteract this tendency.

Figure of Merit

Craig [2] showed that the quality of a DTG can be represented by a *figure of merit* that relates the rotor inertias to the gimbal inertias

$$F_m = \frac{C}{A_g + B_g - C_g} \tag{9.5}$$

The fourth operating condition (listed earlier) would give infinite F_m if it were satisfied. In practice, one of the DTG designer's goals is to make the gimbal as small as possible, because the rotor size is set by the necessary drift performance and the available space.

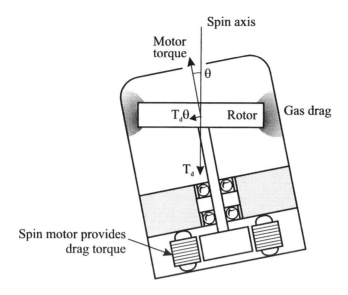

Figure 9.5. Losses causing low time constant.

Damping and Time Constant

Rotor damping can come from energy lost to the rotor in flexure hysteresis, from gas drag, from gas pumping between the rotor and the case or any shaft-fixed component like an angular stop plate, from magnetic leakage from the torquer magnets on the rotor (discussed later), and from windage drags [4]. In Figure 9.5, drag torque T_d acts along the rotor spin axis and is provided by motor torque T_d along the motor-shaft axis. The resultant torque $T_d\theta$ about the input axis causes the rotor to follow the case motion.

To demonstrate this, one can run a DTG open loop and move the gyro case suddenly through a small angle, say $0.5°$. The pickoff will indicate the $0.5°$ displacement, which, in an ideal free gyro, should remain constant forever. In practice the angle exponentially drifts back to zero, and the time it takes to return to $1/e$ (63%) of the initial displacement is the *time constant*. This parameter is an important contributor to gyro quality, as we shall see later.

Biases Due to Damping and Mistuning

In the following we will use terminology similar to that in Chapter 7 extended to two sensing axes. Drift coefficients in the model equation will be D_{xa} referring to a drift on the x axis due to an input along a, D_{ya} for a drift on the y axis, and so on.

What penalty do we pay if the gyro is not operated exactly at tuned speed, either because we did not manufacture it exactly right (for there must be some tolerance for economic assembly), or because it has changed since we built it? Also, the flexure stiffnesses K_x, K_y are proportional to their material's elastic modulus, and the elastic modulus of most flexure materials depends on temperature, so we must either tolerate or compensate that, at least.

A pair of coupled equations describe the gyro drift as a function of time constant, figure of merit, mistuning, and rotor deflection from the mechanical null (the position where the gimbal does not flutter) [2]. The drifts on the x- and y-axes are[1]

$$D_x = \frac{1}{\tau}\phi_x + \frac{dN}{F_m}\phi_y$$

$$D_y = \frac{1}{\tau}\phi_y - \frac{dN}{F_m}\phi_x$$

(9.6)

where
$\phi_{x,y}$ = rotor deflection components
τ = time constant
dN = error in tuning

Because these equations predict a drift proportional to angle, the coefficients are referred to as *elastic restraints*. The first is the *time constant restraint*, D_{xox} or D_{yoy}, and the second is the *mistuning restraint*, D_{xoy} or D_{yox}. If we have a DTG that spins at 12,000 rpm (1256 rad/s) that is mistuned by 1% and has $F_m = 50$ (quite reasonable), the mistuning restraint per arc-sec of pickoff offset is

$$D_{xoy} = D_{yox} = 0.25 \text{ arc-sec/s, or deg/h}$$

If the time constant is 25 s (another reasonable value), the drift per arc-sec pickoff offset is

$$D_{xox} = D_{yoy} = 0.04 \text{ deg/h}$$

These numbers illustrate how necessary it is to design the pickoff so that it has x- and y-axis nulls stable to better than 1 arc-sec. The gyro must run very close to its tuned speed if it is to have good navigation performance; it is a free gyro only because it cancels one large spring torque with another, so it is not surprising that the two torques must be very stable.

---

[1] Note that there is a sign error in these equations in Craig's paper.

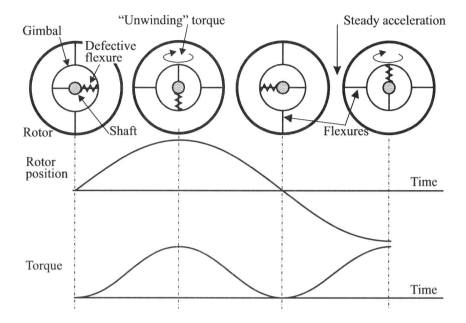

Figure 9.6. Quadrature mass unbalance (*g*-sensitivity).

Quadrature Mass Unbalance

The DTG has mass unbalance *g*-sensitivities just as the SDFG does, caused by the rotor center of mass being offset axially from the flexure plane. An acceleration along the x-axis causes an inertia torque about the y-axis, which causes a gyroscopic drift about the x-axis. This normal mass unbalance coefficient is written as D_{xx}, and its y-axis partner is D_{yy}.

Flexures can generate a torque when loaded axially, in the same way as a piece of rope or a coil spring twists when pulled. This effect leads to a *g*-sensitive drift about the opposite axis to that along which the acceleration acts, represented by drift coefficients D_{xy} and D_{yx}, called *quadrature mass unbalances*.

They arise as shown in Figure 9.6, which shows a rotor positioned so that a steady acceleration, applied to the shaft, acts in the plane of the rotor. We assume that only one flexure of the four has the property that it twists when loaded, although the net effect will be the sum of the drifts from all the flexures. In the leftmost view, the rotor position is such that the rogue flexure is not loaded, so there is no torque generated. A quarter-turn later that flexure is in compression and we assume that the torque has the sense shown. At the next quarter-turn there is no torque, and at the quarter-turn following that, the torque is again generated. Because the flexure is now in tension the torque is in the opposite sense. But the rotor position is now reversed, leading to the unwinding torque acting on the rotor in the same sense as before.

Therefore, over a full revolution there is a nonzero mean torque. By inspection you can see that the acceleration along x causes a torque about x and thus a drift about y. Fortunately these drift coefficients are stable and readily compensated in a navigation system.

Synchronous Vibration Rectification Errors

The DTG is sensitive to vibrations at particular integer multiples (n) of the spin speed, N. These are:

n = 1   1N axial vibration couples with rotor radial unbalance, just as we explained in Chapter 7 for the SDF gyro (D_{xv1}, D_{yv1}),

n = 2   2N angular vibration about the IA interacts with the gimbal angular momentum and rectifies to a fixed drift.

2N radial (linear) vibration couples with gimbal pendulosity, that is, the separation between the axial position of the gimbal center of mass and the inner flexure axis.

2N radial vibration couples to the rotor mass and any separation between the inner and outer flexure axes. This torque is zero if the flexures are coplanar.

To estimate the magnitudes of these errors let us assume a gyro with the following parameters:

m = rotor mass = 20 g
H = 20,000 dyn·cm·s
N = spin speed = 24,000 rpm
F_m = 25

Axial Vibration at 1N

If the axial 1N vibration has amplitude a_o = 1 µm, so that the axial acceleration is 630 cm/s², and the radial imbalance of the rotating assembly (rotor, flexures, gimbal) is r = 1 µm, then the rectified drift will have an amplitude

$$D_{xv1} = \tfrac{1}{2}ma_o r/H = 6.3 \text{ deg/h}$$

The position of the output drift between the x- and y-axes depends on the phase of the vibration with respect to the position of the rotating imbalance. For rectification the vibration frequency must be *exactly* N; such a forcing function will practically

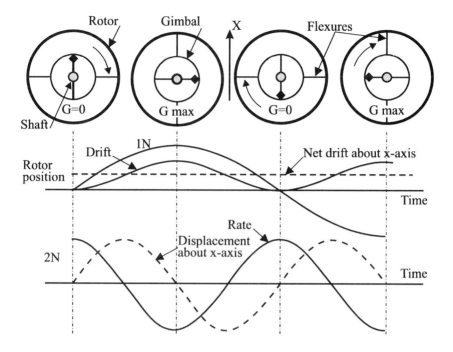

Figure 9.7. DTG angular 2N error.

never arise from environmental vibration but will come from another gyro operating at the same speed nearby. The phasing might vary from run to run as the rotors synchronize in different positions each time, or the transmissibility of the structure might vary and change the vibration amplitude, so this error cannot be compensated by a factory calibration.

One can run two DTGs in a system at slightly different speeds to avoid this error, but in any case one should radially balance the rotor during gyro assembly to as tight a specification as possible. Then one gyro will not generate a disturbing vibration, and the other will not respond to it.

Angular Vibration at 2N

The 2N angular error is more serious than the 1N error. It is generated as depicted in Figure 9.7, which shows "stick" sketches of a DTG wheel in various stages of rotation. Let us again assume that the inertias of the rotor are large compared with those of the gimbal.

At step 1, the angular input has zero amplitude and the rate (although it is at its maximum) is not applied to the gimbal. Thus the gyroscopic torque G on the gimbal is zero. At step 2, a quarter of a revolution later, the input rate is again maximum and the angular position is zero. The angular rate is now transmitted to the gimbal because the flexures are very stiff in bending; therefore, the gimbal tries to precess, generating torque G. As the rotor is coupled to the gimbal at this time by the outer flexure in bending, the rotor has a torque acting on it in the direction shown. Step 3 is the same as step 1 in that there is no torque on the rotor, but step 4 repeats step 2, giving a torque on the rotor in the same direction as well. In this way the input motion is rectified by the rotating gimbal.

If the gimbal is rotated 90° with respect to the input displacement, steps 1 and 3 give a torque, whereas 2 and 4 do not. Therefore the magnitude and sign of the generated torque depend on the phasing of the 2N input with respect to the flexure axis. As before, the drive frequency must be exactly 2N to be rectified.

Craig [2] shows that the 2N drift has a maximum of

$$D_{xv3} \text{ (or } D_{yv3}) = N/4F_m \text{ per arc-sec at 2N}$$

For our model gyro, $D_{xv3} = 25$ °/h/arc-sec, such an enormous value that we must examine its implications.

It is not easy to generate 2N vibrations. One gyro can shake another at 2N if there is an anisoelastic mounting between them, but it is not likely. Self-generated 2N is the usual source. But there are not many sources of 2N vibration in a gyro; the bearings do not generate it in any direct sense. The planetary gear action of the balls in the races generates noise at many frequencies as balls run over irregularities in the inner and outer races. The retainer that locates the balls in the bearing carries out its own independent motion, traveling in cycloidal orbits (sometimes in an unstable whirl). But these noises are all at noninteger multiples of N.

One source of 2N is the radial magnetic force from interaction of an out-of-round spin motor stator with an eccentric drive rotor on the end of the shaft. As the bearings have finite radial compliance, a rotating force at 2N in the motor is converted into a 2N angular input.

Whatever the source of the 2N input, it causes varying bias in the gyro if its magnitude is not constant with respect to inertial space, so that any change in the compliance of the gyro mounting in the system (perhaps due to the hardening of an elastomer gasket at low temperatures) will change the 2N drive at the rotor and give a bias change.

Wide Band Vibration Rectification Errors

These errors occur in the frequency band from 0 to 2kHz or so and comprise the anisoelasticity, anisoinertia, and pseudoconing errors. The first two were already described in Chapter 7, whereas the third is new; it is not found in single-axis gyros.

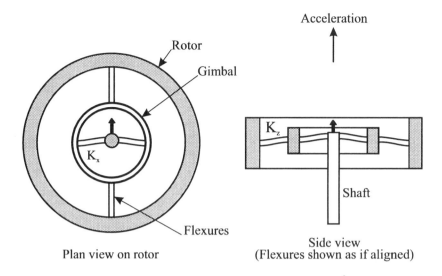

Figure 9.8. DTG flexure isoelasticity.

Anisoelasticity

The SDF gyro suffered from anisoelasticity if the wheel bearings and shaft mounting had different compliances along the spin and input axes (Chapter 7). The DTG can also have anisoelastic drift, but not from the spin bearings, because they do not control the position of the rotor relative to the suspension axis, the only axis about which drift torques can act. DTG anisoelasticity comes from the flexures, if the axial stiffnesses do not match the radial stiffnesses. The stiffness from the rotor to the case in the axial direction comes from summing the outer and inner flexure compliances and doubling that stiffness (as there are two of these in parallel), as shown in Figure 9.8. Let K_z be the stiffness of an inner and an outer flexure along the spin axis. Then the axial stiffness is

$$K_a = 2(K_z/2) = K_z$$

and the radial stiffness is that due to only one pair of flexures. Let K_x be the stiffness of each flexure in the plane of the rotor. Then the radial stiffness is

$$K_r = 2K_x$$

For isoelasticity, $K_r = K_a$, i.e., $K_z = 2K_x$, so that each flexure must have unequal radial and axial stiffnesses; the anisoelastic drift coefficient [Equation (2.4)] for the x-axis is

$$D_{xxz} = \tfrac{1}{2}m^2(C_x - C_z)$$

where C_x, C_z = net x and z rotor support compliances.

Not only must the flexure assembly be isoelastic, but the flexure stiffnesses must be high enough that the resonant frequencies of the rotor in the radial and axial directions are outside the range of environmental vibrations. That is usually 2kHz in airborne systems; because of the low damping designed into the DTG, any resonances below this frequency can have a high Q and build up damaging rotor oscillations. Because the flexures are designed to have as low a stiffness as possible, environmental vibrations can fatigue the flexure material and cause premature failure—the flexures crack and break. To avoid fatigue, the stress in the flexure must be kept below a limit specified for its material.

Designing a flexure and gimbal assembly is as much art as science. To keep it tuned, isoelastic, with high resonant frequencies and stresses inside the fatigue limit, with a high figure of merit, stable over time and temperature, and, of course, inexpensive to make, demands the highest mechanical engineering skill.

Anisoinertia

The anisoinertia error is frequency sensitive (as in the SDFG), but in a different way [5]. When the DTG experiences simultaneous rotation rates about one input axis and the spin axis, at frequencies well below the spin motor hunt frequency, the wheel angular momentum is altered by the input rate

$$H_z = C(N + \Omega_z)$$

where Ω_z = rotation rate component about the SA. The gyro angular momentum due to wheel spin alone is $H = CN$. The y-axis servo cages the rotor to the case rotation about the y-axis, and the rotor therefore has a transverse angular momentum

$$H_y = A\Omega_y$$

where
  A = rotor transverse inertia
  Ω_y = rate about y-axis

The x-servo must provide a torque

$$T_x = H_z\Omega_y - H_y\Omega_z$$
$$= C(N + \Omega_z)\Omega_y - A\Omega_y\Omega_z$$
$$= H\Omega_y + (C - A)\Omega_y\Omega_z$$

The second term is the anisoinertia error term and has a drift coefficient

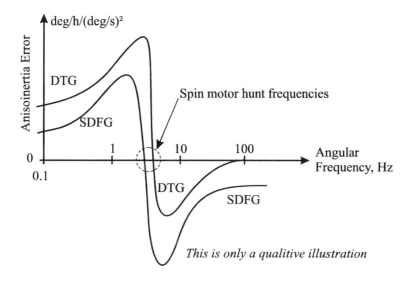

Figure 9.9. Comparison of DTG and SDFG dynamic anisoinertias.

$$D_{xi} = (C - A)/H$$

The only way that this error can be zero is if C = A, a condition met by a spherical rotor; normally A < C. However, increasing A has the contrary effect of increasing the pseudoconing error (explained later).

If the input is oscillating, the component along the spin axis is coupled through the spin motor to the shaft and rotor. Assuming that the motor is a hysteresis type, at frequencies below the motor hunt frequency, the disturbance is transmitted to the rotor, and at the hunt frequency the motor resonance Q-factor will amplify Ω_z. Above the hunt resonance the motor acts as an isolator, no disturbing rate passes through to the rotor, and the anisoinertia error drops to zero. Compare this to the behavior of the SDFG, whose anisoinertia error is still finite at high frequencies (Chapter 7); the two types of gyro are compared qualitatively in Figure 9.9.

Pseudoconing

The rotor transverse inertia is the cause of another possible dynamic error. Suppose that a DTG with two ideal servos rebalancing it experiences an input along its x-axis given by

$$(d\phi/dt)_x = (d\phi/dt)_o \sin \omega t$$

The torques provided by the servos driving the torquers are

$$T_x = A(d^2\phi/dt^2) \qquad \text{Newton's Second Law}$$
$$\quad = A\omega(d\phi/dt)_o \cos \omega t$$

and

$$T_y = -H(d\phi/dt)_x \qquad \text{Gyroscopic term}$$
$$\quad = -H(d\phi/dt)_o \sin \omega t$$

The navigation computer receives two signals from the gyro 90° apart in phase and assumes that the vehicle axis is coning, although it really is not. Unless there are some other data that say that there is no real motion about the y-axis, this *pseudoconing* error cannot be removed.

The Pickoff and Torquer for a DTG

The most common pickoff for a DTG is the electromagnetic variable transformer. Other pickoff types using windings on the torquer, Hall effect sensors, or optical devices have also been used. The electromagnetic pickoff uses a magnetic circuit formed from a case-mounted core and a plate attached to the rotor. The core and rotor plate are made of a magnetically permeable material such as ferrite or nickel iron; actually, the entire rotor is often made of nickel iron, except for the permanent magnets, which we will discuss when we consider the torquer. As shown in Figure 9.10, the core carries primary and secondary coils, and the primary carries a current that establishes a magnetic flux around the circuit whose magnitude depends on the gap between the core and rotor plate. The flux links with a secondary coil and induces an electromotive force in it. Two pickoff core/coil assemblies are arranged on opposite sides of the rotor so that the gap between the core and rotor plate varies in the opposite sense as the rotor tilts, and the e.m.f.s generated in the secondaries are connected in opposition. The voltage across a high impedance load is the pickoff output.

The electrical null, where $\phi_x = \phi_y = 0$, is adjusted to coincide with the mechanical null by shunting the larger of the two outputs. The nulls must be stable over time and temperature, so the pickoff output must be stable in amplitude and phase; if not, we get the elastic restraint errors described by Equations (9.6).

There is a residual voltage at null caused by harmonics from the magnetic circuits (perhaps caused by eddy currents and hysteresis in the cores), which are not identical in amplitude and phase and so do not cancel when the fundamental signals cancel. The residual must be low so that the null angle uncertainty is small.

In the SDF gyro (Chapter 7) the torquer used case-fixed magnets and coils on the float, and a current in the coils generated a torque around the output axis (OA) by interaction with the radial magnetic flux. While it is straightforward to run current to the float through the flexleads, it is not feasible to run current to the spinning DTG rotor through slip rings and across the flexures to the rotor; that would add damping, reduce τ, and cause drift [Equation (9.6)]. So we put the magnets on the rotor and the coils on the case, and now we have to shield against

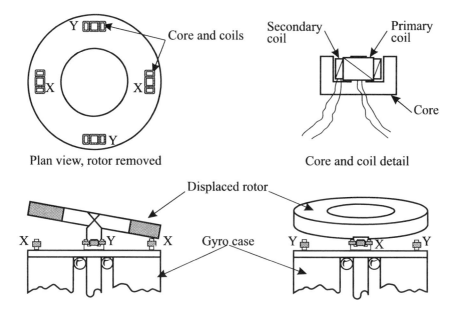

Figure 9.10. Typical DTG pickoff.

the possibility that external magnetic fields (like that of the earth) can interact with the rotor fluxes and cause the rotor to drift.

The DTG torquer is more difficult to design than the SDFG torquer, as it requires a more careful optimization of the magnetic circuit and coils. The procedure is outlined later. Figure 9.11 shows a basic DTG permanent magnet torquer with a single magnet ring driving flux across a gap in which the torquer coil is located. Current running circumferentially in the coil interacts with the radial flux and generates a torque on the rotor

$$T = Bilnr \tag{9.7}$$

where
    B = magnetic flux density
    i = coil current
    l = length of conductor in the flux
    n = number of coil turns
    r = radius of action of coil

To generate torques around both x- and y-axes, we need four coils, each spanning a quadrant, with the opposite pairs connected so that their torques add.

The torque required is set by the gyro's maximum operating rate, Ω_{max}

$$T_{max} = H\Omega_{max} \tag{Equation 6.2}$$

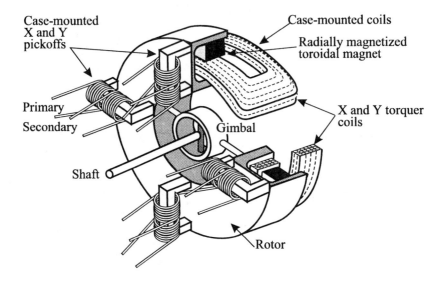

Figure 9.11. DTG pickoffs and torquers.

The power dissipated in the torquer is

$$P = i^2 R \qquad (9.8)$$

The coil resistance is

$$R = \rho L/a = \rho n L_t/a \qquad (9.9)$$

where
   ρ = wire resistivity
   L = length of wire
   L_t = length of the mean turn
   a = wire cross-sectional area

Combining (9.7)-(9.9) gives the power dissipated

$$P = \frac{\rho T^2 L_t}{B^2 l^2 r^2 (na)}$$

As (na) is approximately constant, the power dissipated does not depend on the size of the wire, all other parameters being constant, and provided the wire fills the

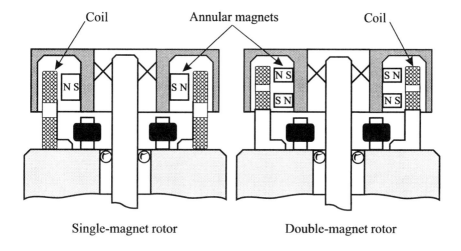

Figure 9.12. DTG 1- and 2-magnet ring torquers.

available space. But the maximum power that can be dissipated depends on the area available for the winding, because if it is too small the coil will burn out; there is a maximum safe current density for the wire in a coil.

There is an optimum torquer design for the DTG, just as there was for the pendulous accelerometer forcer (Chapter 4). If the rotor has fixed volume, as it usually does in real life, should you use a thin magnet giving a low gap flux density but allowing you to put a large coil in the gap, or should you use a thick magnet giving a high flux density but a small coil?

The bigger magnet will need thicker iron sections in the rotor so that the higher flux densities do not saturate the iron, because saturating the iron causes large gyro biases as the flux leaks out of the rotor and interacts with the case and gimbal. Thicker sections add to the rotor inertias, the angular momentum, and thus to the torque demanded.

One can use computer modeling to design the rotor iron so that it is everywhere below saturation but has the lowest inertia possible. Then an optimization program can proportion the magnet and coil so that the coil current is below the overheating limit, giving maximum torque per unit power dissipated in the coils. For higher performance (but higher cost) one can use two ring magnets and both sides of the coil, as shown in Figure 9.12.

The magnets are commonly made from rare earth materials such as samarium cobalt. Their generated fluxes depend on temperature (typically 0.04%/°C for SmCo), which can be modeled if you know the magnet temperature, although that is not easily obtained as the magnet is spinning inside the rotor.

The DTG Model Equation

The drift coefficients described earlier can be combined into a pair of equations (one for the x-axis and one for the y-axis) that describe the gyro performance. We will write the equation for the x-axis:

$S_x/K_x = \Omega_x$	Inertial rate
$+ B_x$	Fixed bias
$+ D_{xx}a_x$	Normal g-sensitivity
$+ D_{xy}a_y$	Quadrature g-sensitivity
$+ D_{xz}a_z$	"Dump" g-sensitivity
$+ D_{xox}\phi_x$	Offset sensitivity, x
$+ D_{xoy}\phi_y$	Offset sensitivity, y
	Synchronous vibration sensitivities:
$+ D_{xv1}a_{1Nz}$	1N axial
$+ D_{xv2}a_{2Nxy}$	2N radial
$+ D_{xv3}(d\phi/dt)_{2Nxy}$	2N angular
$+ D_{xxz}a_za_x$	Anisoelasticity
$+ D_{xi}(d\phi/dt)_x(d\phi/dt)_z(\omega)$	Frequency-sensitive anisoinertia

The fixed bias may be due to improper null setting or mistuning or to some residual torque on the rotor from magnetic leakage, for example, [6]. The term D_{xz} (the "dump" term) may be due to the variation of bearing-generated vibration with bearing load, which is changed when the gyro is turned from SA-up to SA-down, the drift torque being generated from the rectification of the internal vibration. The anisoinertia term D_{xi} is described in terms of the rates of change of the case angles ϕ_x and ϕ_z, assumed to be periodic with angular frequency ω.

Conclusion

The flexures and gimbals are the heart of the DTG, and their design is more complex than we can explore here. There are good descriptions of the tradeoffs involved in papers by Haberland, Karnick, and others [7–11].

   DTGs are cost-effective where rate information is needed about only two axes, as in a missile seeker head; they are small and rugged. Because they are dry, they provide good performance over wider temperature ranges than the SDFG, even though the SDFG has fewer sources of error and can have lower drift. The electronics servos for a DTG are more complicated than those for a SDFG, because of the DTG's low damping, cross-coupling, and nutation. Joos [12] compares these two gyros in greater detail.

References

1. Howe, E.W., P.H. Savet, "The dynamically tuned free rotor gyro," Control Engineering, pp. 67–72, June 1964.
2. Craig, R.J.G., "Theory of operation of an elastically supported tuned gyroscope," and "Theory of errors of a multigimbal, elastically supported tuned gyroscope," IEEE Trans. on Aerospace and Electronic Systems, AES-8, 3, pp. 280–297, May 1972.
3. Willems, P.Y. (Ed.), *Gyrodynamics*, Springer-Verlag, New York, 1974. Includes: Maunder, L., "Some designs and features of dynamically tuned gyroscopes"; Fogarasy, A.A., "A contribution to the dynamics of a spring restrained Hooke's joint gyroscope"; Lawrence, A.W., "A simulation of a dynamically tuned gyroscope."
4. Karnik, H., "Experience based upon experimental dry tuned gyros," DGON Symposium Gyro Technology, Stuttgart, 1979. Contributes useful insights into the causes of low time constant.
5. Craig, R.J.G., "Dynamically tuned gyros in strapdown systems," AGARD Conference on Inertial Navigation Computers and Systems, Florence, Italy, Oct. 1972.
6. Albrecht, W.G., "Errors of DTGs on acceleration induced mistuning and rotor axis tilt," DGON Symposium Gyro Technology, 1983.
7. Haberland, R., "Some design criteria of elastic universal joints for dry tuned gyroscopes," DGON Symposium Gyro Technology, Stuttgart, 1977.
8. Carroll, R., "Dynamically tuned gyroscope," in Ragan, R.R. (Ed.) "Inertial technology for the future," IEEE Trans. on Aerospace and Electronic Systems AES-20, 4, pp. 414–444, July 1984.
9. Beardmore, G., "The design and development of a novel strapdown DTG incorporating a gas bearing and fabricated flexure hinge," DGON Symposium Gyro Technology, Stuttgart, 1984.
10. Shimoni, U., "Tamam miniature gyro," DGON Symposium Gyro Technology, Stuttgart, 1983.
11. Karnick, H., "Effects of hinge improvements on design and production cost of a dynamically tuned gyro," DGON Symposium Gyro Technology, Stuttgart, 1978.
12. Joos, D.K., "Comparison of typical gyro errors for strapdown applications," DGON Symposium Gyro Technology, 1977.

10
Vibrating Gyroscopes

Gyroscopes would be more reliable and less expensive if they had neither spinning wheels nor flotation fluids. Single-degree-of-freedom gyro (SDFG) wheel bearings must be stiff and isoelastic, and for both SDFGs and dynamically tuned gyros (DTGs) the bearings must be noiseless, as low friction as possible, and must start at low temperatures. Gas bearings cost more, and flotation fluids can require temperature control, damping compensation, and scrupulously clean assembly. By the 1960s every reasonable path around these problems had been tried.

Engineers began to try alternatives to the wheel, using vibrating rather than rotating bodies to provide gyroscopic torques from the Coriolis acceleration (Chapter 6). After all, nature has provided one flying insect, *diptera*, with tuning fork gyros (*halteres*) for its flight control! Most gyro engineers were developing vibrating gyros in 1965, based on vibrating strings and tuning forks; most had discovered their problems and abandoned the field by 1970.

However, three separate innovations have revived this field. The first is a new concept based on a very old idea, the second, an old idea animated by new technology, and the third, a novel invention.

The first comes from 1890, uses the theory of ringing bells, and is represented by Delco's Hemispherical Resonator Gyro (HRG) and by the British General Electric Company's START (Solid State Angular Rate Transducer). The second is illustrated by the Systron Donner Quartz Rate Sensor (QRS), a micromachined crystalline quartz tuning fork gyro, and by the Draper Labs' micromachined silicon tuning fork gyro. Pittman provided a third innovation in the 1960s; he rotated piezoelectric crystals so that they vibrated when subjected to acceleration and vibration. Two versions of this gyro exist; one, the British Aerospace Dual Axis Rate Transducer (DART), and the other, offered in different designs by Kearfott and Rockwell Collins, the multisensor.

In this chapter we will describe the principles behind these three different ways of using the Coriolis acceleration, using particular designs as case studies.

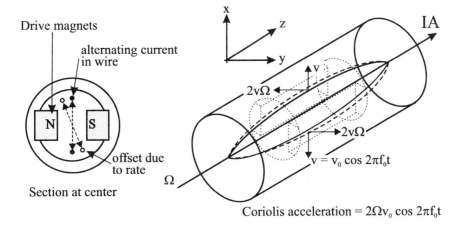

Figure 10.1. The vibrating string gyro.

The Vibrating String Gyro

A vibrating string oscillates in a plane, and if the string's supports turn about the string axis, the plane of vibration tends to stay fixed in space, even though the string rotates with its supports. The string acts like a Foucault pendulum, although the string has a minuscule peak momentum compared to a pendulum.

Figure 10.1 shows a sketch of a vibrating string gyro. The (x,y,z) coordinate set has the string lying along the z-axis (the IA), vibrating in the xz-plane with instantaneous velocity $v = v_0 \cos 2\pi f_0 t$. A pickoff is positioned to measure the component of vibration along the y-axis; under no rotation, it has no output. Under a constant rate Ω about the string axis, a Coriolis acceleration [Equation (6.5)] of $2v\Omega$ causes the vibration to be coupled into the y-axis, whose amplitude is proportional to Ω.

Synchronous case vibrations along the pickoff axis are interpreted as inputs, so that the string natural frequency and the driving frequency (ideally the same) should be an order of magnitude higher than the environmental vibrations. There is usually little environmental noise above 2 kHz, implying the need for a string frequency $f_0 > 2$ kHz.

Quick [1] has described the obstacles to be overcome in achieving useful performance from such a gyro. A string differs from a bar in that the string is infinitely flexible—it does not transmit bending moments. But at the ends of the string, at the attachment points, there will be regions where the curvature is high, and the bending moment becomes important, so that if there is elastic anisotropy or geometric asymmetry there, the effective string length varies with angular position. This anisoelasticity causes large drift errors, partly avoided by operating the gyro at null on an inertial platform. Bias torques also arise from damping asymmetry in gas drag and end attachment losses. The error torques tend to be so

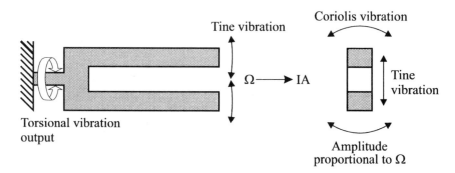

Figure 10.2. The tuning fork gyro.

large compared with the string momentum that no one has successfully marketed a vibrating string gyro.

The Tuning Fork Gyro

A single piezoelectric bar can make a gyro similar to a vibrating string [2], but it is difficult to establish inertially stable end mounting conditions for a bar. However, as described for the vibrating beam accelerometer (Chapter 5), a balanced system like a tuning fork, made from two bars oscillating anti-phase, has no net motion at their junction and can be mounted there.

The fork tines vibrate in their plane with a sinusoidally varying angular momentum due to their simple harmonic tip velocity. If the vibrating fork is rotated at a steady speed about its axis as shown in Figure 10.2, the Coriolis acceleration generates a sinusoidally varying precession about that axis, whose amplitude is proportional to the input rate [3]. Ignoring damping and inertia torques, the instantaneous output amplitude for two tines is

$$a = 4v\Omega/K$$

where
v = tine velocity = $v_o \sin \omega_f t$
ω_f = tine frequency
Ω = steady input rate
K = stem torsional stiffness constant

The angular momentum of a vibrating tine is so small compared to that of a spinning wheel that the gyroscopic torques generated can easily be overwhelmed by disturbing torques. The gyro sensitivity can be increased if the fork stem is tuned to torsional resonance at the fork's natural frequency.

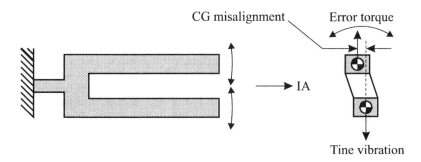

Figure 10.3. Tuning fork gyro mass shift error.

Unlike the (unbalanced) vibrating string, the balanced oscillator is not sensitive to linear vibration. But it does respond to angular vibrations about the stem axis [4], and again the tine frequency should be an order higher than the environmental vibrations.

Two problems in particular stalled tuning fork development for more than a decade, frustrating the efforts of those working with metal forks. One is that the bending and torsional elastic moduli of fork materials vary differently over temperature, so that the two resonances do not track, and the other is the bias instability caused by the lateral displacement of the tine mass centers. The mass instability problem can be explained with reference to Figure 10.3, which shows the tines viewed down their length, towards the base and shows the tine centers of mass. If these centers of mass are not precisely in the plane of vibration, their inertia forces produce a vibrating torque about the stem, just as a Coriolis torque does. The two motions are out of phase, but the noise created by the mass center shift (due probably to differential tine bending over temperature) swamps the gyroscopic torque. Hunt and Hobbs [5] describe the difficulties they encountered in making a tuning fork gyro.

In the same way that metal vibrating string accelerometers gave way to the vibrating beam accelerometer (Chapter 5), changing to crystalline quartz tuning forks has relieved some of these difficulties. Digital wrist watches owe their accurate timekeeping to the time base that each carries, made from a quartz crystal oscillating at a precisely known and stable frequency. A large effort has been invested in the technology of quartz oscillators, stimulated by the vast market for digital watches, with the result that quartz tuning fork crystals only a few millimeters long and wide can be made inexpensively using photolithography to define the mechanical structure in the quartz. This new micromachining capability, similar to silicon micromachining (Chapter 4), allows the quartz rate sensor tuning fork gyro to be made cheaply, and they are attractive to users because they are small, inexpensive, and take only tens of milliwatts of power.

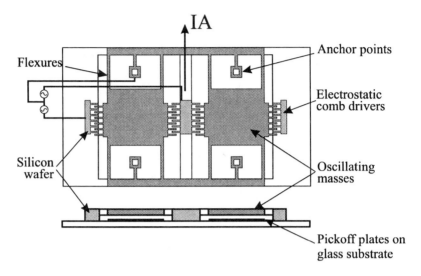

Figure 10.4. Planar micromachined silicon tuning fork gyro.

The Micromachined Silicon Tuning Fork Gyro

Draper Labs, in association with Rockwell, has made a micromachined silicon planar tuning fork gyro as shown in Figure 10.4, on a die only 1 mm on a side. This gyro performed at about the 20-deg/h resolution level; it is very rugged, having withstood 120,000-*g* shocks.

The gyro is made from an etched silicon piece anodically bonded to a glass substrate [6, 7]. The glass carries the electrostatic driving and capacitive pickoff plates; the silicon piece attaches at the anchor points, leaving most of the silicon to float. The floating "tines" are electrostatically driven by the comb drivers, also etched out of silicon, whose stators are fixed to the glass substrate. These drive the tines so that they oscillate with velocities 180° out of phase with one another. Figure 10.5 shows a micrograph of a Draper gyro.

Under rotation Ω about the input axis (which is in the plane of the tines), the Coriolis forces will make one tine rise and the other drop out of this plane, as shown in Figure 10.6. The capacitive pickoff plates sense the out-of-plane oscillation amplitude, a:

$$a = 2a_0\Omega \sin\omega t$$

In an open-loop gyro this Coriolis signal, demodulated at the tine frequency ω, will be the output, linear in rate if the tine amplitude a_0 is stable. If preferred, a servo can null the pickoff output by applying voltages to "forcer" plates on the

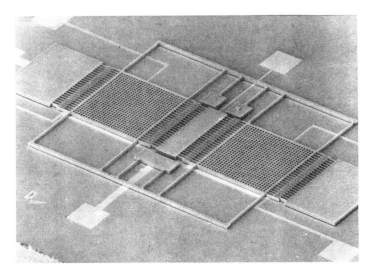

Figure 10.5. SEM photo of micromachined TFG. (Courtesy of Draper Lab.)

glass substrate. One would expect the added complexity to give better linearity and lower sensitivity to orthogonal rates (cross-coupling).

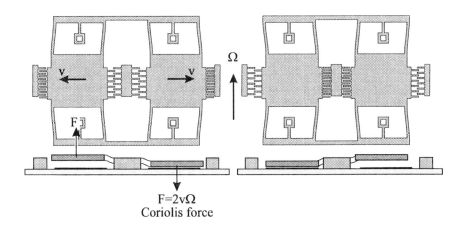

$$F=2v\Omega$$
Coriolis force

Figure 10.6. Operation of the planar tuning fork gyro.

Vibrating Shell Gyros

Tuning fork gyros depend on the rotation-induced transfer of energy between two different vibration modes, tine bending and stem torsion. This can give the gyro large temperature sensitivities because of the different temperature sensitivities of the mode natural frequencies. In contrast, vibrating shell gyros transfer energy between two identical vibration modes, thus avoiding this problem.

Vibrating shell gyros have a bell-like structure. They may be shaped like a wineglass, like the Hemispherical Resonator Gyro (HRG), or they may be cylindrical, like the Solid State Angular Rate Transducer (START). Both operate on the same principle, that Coriolis forces produced by rotation cause a transfer of energy between two of the gyro's modes of vibration.

They are constructed differently for use in differently performing systems. The HRG [8] has demonstrated drift lower than 0.005 deg/h, scale factor stability of 0.02 ppm, and readout noise of 0.02 arc-sec. The HRG is made in fused quartz and has capacitive drivers and pickoffs, whereas the START is made in metal and has piezoelectric drivers and pickoffs. START is much cheaper than HRG but is also much less accurate, having a resolution of about 0.02 deg/s and scale factor of 0.3%. But before we go further with these descriptions, here's a word about their underlying principle.

Lord Rayleigh, in his "Theory of Sound" (1894), analyzed the vibration patterns of plates and shells and referred to work published by a colleague, G.H. Bryan, in 1890 [9] concerning the effects of rotation on the nodal patterns of vibrating shells. Bryan wrote:

> If we select a wine-glass which when struck gives, under ordinary circumstances, a pure and continuous tone, we shall on twisting it round hear beats, thus showing that the nodal meridians do not remain fixed in space. And if the observer will turn himself rapidly round, holding the vibrating glass all the time, beats will again be heard, showing that the nodal meridians do not rotate with the same angular velocity as the glass and observer.

Bryan concluded that the nodal angular velocity was about three-fifths that of the glass; we can therefore measure the precession of the nodes relative to an axis fixed to the shell and derive rotation rate from it.

A thin, long, cylindrical shell vibrates in modes where the radial and tangential displacements vary sinusoidally around the perimeter. The number of nodal diameters describes the mode; the fundamental mode has $n = 2$, and the deflected perimeter is an elliptical standing wave. A standing wave can be represented by two counterpropagating traveling waves, with the same amplitude, moving at the same speed. Rotation about the axis induces Coriolis accelerations, which make the traveling wave speeds unequal. The radial component of the perimeter displacement [4] is

$$r = 2An \cos [n\theta + 2n\Omega t/(n^2+1)] \cos \omega_n t$$

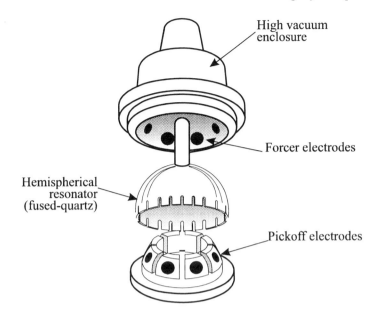

Figure 10.7. The hemispherical resonator gyro.

where
  n = mode number
  θ = cylindrical coordinate
  A = amplitude
  ω_n = cylinder natural frequency

which shows that the elliptical displacement pattern continuously rotates relative to the cylinder, in the opposite direction to Ω, at an angular velocity $2n\Omega/(n^2+1)$. When the shell rotates through angle θ, the nodal pattern moves through θ_2, for example,

$$\theta_2 = \theta(n-1)^2/(n^2+1) = k\theta \tag{10.1}$$

For $n = 2$, $\theta_2 = 0.2\theta$.

The Hemispherical Resonator Gyro

The Hemispherical Resonator Gyro (HRG) is shown simplified in Figure 10.7. The vibrating shell is a fused-quartz hemisphere supported by a stem along a diameter, like a wineglass with its stem continued into the bowl. The shell is machined so that it is as uniform as possible in wall thickness and then dynamically balanced by removing material from around the rim, to compensate for material inhomogeneities and residual machining errors. It is coated with a thin layer of chromium to make it electrically conductive.

The HRG shell is electrostatically excited at the shell's fundamental natural frequency by an AC signal applied to case-fixed electrodes. A servo system drives the shell to resonance and maintains the oscillation amplitude constant. To avoid coupling the node pattern to the case, the enclosure is evacuated as well as possible, and getters remove the last traces of gases and vapors. As the internal damping of the quartz is so low and the enclosure is evacuated, little energy needs to be supplied to maintain resonance. With fused quartz resonators, Delco's design achieved Q factors [Equation (3.4)] greater than 6×10^6, with time constants of the order of 1000 s, almost 17 min. Consequently it is possible to run the gyro with intermittent power to the shell, for drive energy needs to be supplied to the shell only every 10-15 min.

Following the current trend to the consolidation of the aerospace industry, Litton recently purchased the Delco design.

Scale Factor

Capacitive pickoffs measure the location of the anti-nodes. When the resonator is excited, it takes up a mode with nodes shown in Figure 10.8, and when it rotates about the stem, the anti-nodes precess as shown. During a 90° clockwise (CW) rotation about the stem, the anti-nodes precess about 27° CW; as mentioned earlier for the cylindrical shell [Equation (10.1)], this gain k = 0.3 is a geometric constant for the resonator. Generally,

$$\theta_I = k\theta_C$$

where
θ_I = inertial angular rotation of the anti-nodes
θ_C = case rotation angle

The pickoffs measure the angle between the new line of nodes and the new case position, θ_P. The case's inertial rotation is therefore $\theta_C = \theta_I + \theta_P = [1/(1-k)]\theta_P$, and the scale factor $\theta_P/\theta_C = 1-k$.

If the shell is nonuniform, the nodes tend to rest along an axis of minimum energy; they become "pinned" in place and do not move under small rotations.

Asymmetric Damping Error

The principal drift source of the HRG is asymmetric damping, arising from internal damping inside the resonator material, damping from the interface between the resonator and its metal plating, nonuniformity in the resonator surface finish, and nonuniform clearances.

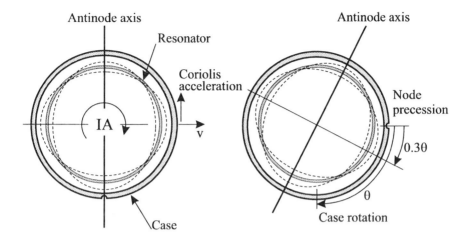

Figure 10.8. Precession of HRG antinodes.

The circumferential variation in damping leads to two different time constants, and the vibration pattern anti-node axes do not coincide (generally) with either normal damping axis. Thus the amplitudes of the vibration components decay at different rates, and eventually one component has died away while the other is still vibrating. This leads to vibration in the "preferred" resonator axis (that with least damping), so that the vibration pattern drifts toward this preferred axis at a rate determined by the damping asymmetry and how far the vibration is from the preferred position.

Better than 0.1 μm surface finish is necessary for acceptable levels of this drift; HRG performance is achieved by compensating the output as a function of readout angle, temperature, and drive voltage.

The temperature coefficient of Young's modulus for fused quartz causes a frequency variation of about 80 ppm/°C in the resonant frequency. The electronics directly measure this frequency, providing a temperature signal for drift modeling.

Loper and Lynch [10, 11] detail HRG performance and error sources and include photographs of the parts. The high performance is achieved by precise machining and high vacuum assembly, making the HRG inherently expensive, although efforts have been made to reduce its cost [12].

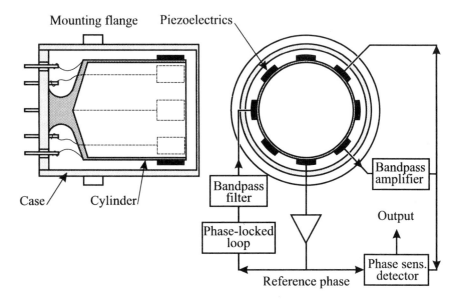

Figure 10.9. The START gyro.

The Vibrating Cylinder (START) Gyro

At the other end of the performance-price scale, the START gyro uses piezoelectric pads to excite a cylindrical shell so that there are nodes at the quadrature points under no rotation. Piezoelectric pads equally spaced between the drive pads act as pickoffs, as shown in Figure 10.9. Pads A and B vibrate the cylinder; they are connected in a phase-locked loop so that it resonates at its natural frequency. Pickoff pads (C) are on a vibration node when the cylinder is not rotating, but when the cylinder rotates about its axis, the nodes precess and the resulting displacement at C is a measure of rate. Demodulating the signal from the piezoelectrics at C with respect to the drive to A provides a DC signal whose amplitude is proportional to rate, and whose sign indicates the sense of the rotation. Fox [13] and Burdess [14] have published detailed analyses of the cylindrical shell gyro.

The cylinder's high Q gives the gyro a narrow bandwidth. But the response can be shaped by appropriate feedback using other piezoelectrics (D), which drive the cylinder with a modified version of the C signal. This loop's frequency response determines the net gyro's natural frequency and damping, which can thus be controlled entirely electronically. GEC has achieved 90 Hz bandwidth and 0.7 of critical damping [15].

The scale factor can be varied widely (by changing one resistor in the electronics) without changing the cylinder, and the range can be scaled to 1000 deg/s; typical linearity is 0.5%. Scale factor temperature sensitivity is due to expansion mismatch between the piezoelectric and cylinder materials; with compensation, START can achieve 5% from – 40 to 80°C.

The signal-to-noise ratio at the C pickoffs determines the threshold, typically 0.03 deg/s. Bias and bias stability depend on the geometric accuracy of the assembly, its stability over temperature and time, and the stability of the electronics circuit parameters. Bias can be trimmed on assembly to ± 2 deg/s; bias-temperature sensitivity is 10 deg/s over – 40 to 80°C, and improvement is being pursued. START typically has g-sensitivity of the order of 0.05 deg/s/g normal to the input. Startup time is in the range from 0.1 to 0.5 s; the drive power is about 10 mW.

The START gyro trades simplicity for performance. The pads are cemented to the shell, and wires bring power to the drivers and take signals from the pickoffs. Both the pads and the wires damp the resonance and reduce its Q. START uses a metal cylinder that has more internal damping than the HRG's fused quartz bell. However, it is a very rugged design; it can withstand gun-firing shocks of greater than 30,000g. Although the START could be useful in smart munitions and hypervelocity projectiles, it does not have the potential for navigation performance.

Koning's vibrating cylinder gyro [16] differs from the START in that it uses a magnetically driven ferromagnetic cylinder with capacitive pickoffs for the express purpose of reducing damping on the cylinder.

The Advantages of Vibrating Shell Gyros

The HRG's high Q gives it a long time constant—it continues to resonate when the power is removed. Thus it "remembers" rotations occurring while it is temporarily unexcited. So do spinning wheel gyros like the SDFG, and to a smaller extent the DTG, but the optical gyros "die" instantly if they lose power. This memory of the angle turned could be useful in guidance systems for missiles that have to operate near nuclear blasts, such as anti-ballistic missiles. START, with lower Q, has a shorter memory, of course.

Vibrating shell gyros are rugged. Their shape gives them much more immunity to external vibration than the simple tuning fork, since there is only weak interaction between the shell vibration and the gyro mounting.

They consume little power, and because they have no moving parts, vibrating shell gyros should have a very long life, provided that a good hermetic seal can be obtained and internal outgassing can be avoided.

The HRG is expensive to manufacture, because its good performance comes from the precise machining of the shell and housing, the high vacuum sealing and gettering, and the electronics. It is about as difficult and expensive to build as an SDF gas bearing floated gyro or a ring laser gyro (Chapter 13). The fact that its output needs to be corrected as a function of position means that it needs the services of a computer. On the other hand, the inexpensive START has rate gyro performance, typically 0.1 deg/s.

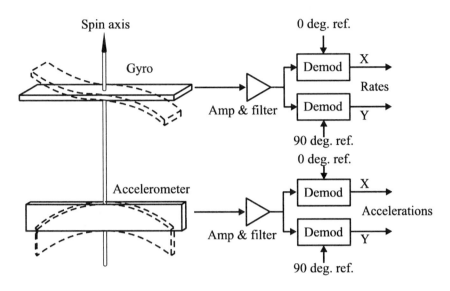

Figure 10.10. Multisensor operation.

The Multisensor Principle and Its Error Sources

A stressed piezoelectric crystal generates a charge, a phenomenon that is used to make accelerometers. Such accelerometers do not have very low thresholds, nor very stable day-to-day bias, so they are not used in inertial navigation systems. However, these problems can be side-stepped if the instrument is designed to work only at a set frequency, because then synchronous demodulation will remove DC uncertainties.

Pittman's innovation [17] was to spin a piezoelectric crystal at a fixed speed and to electronically demodulate the signal at the spin speed to recover the baseband signal. He measured rotation rate by spinning a piezoelectric crystal with its axis at right angles to the accelerometer crystal, so that it sensed the Coriolis acceleration. These crystals are shown in Figure 10.10; a single sensor provides two axes of rate and acceleration and is known as a *multisensor*. To see how the signals are generated, assume first that an acceleration acts in the direction shown in Figure 10.11. In positions 1 and 3 the crystal will bend in an arc, and a charge will be generated. As the crystal spins one-quarter of a turn, to positions 2 and 4, it is aligned with the acceleration along the long side of the crystal, the crystal is straight and the charge drops to zero. A quarter-turn later it is in position 1, and so on. An angular reference generator on the spin axis measures the spin speed and provides a reference pulse that indicates the crystal attitude relative to a datum. These signals provide the reference for the demodulator, which generates a signal proportional to the acceleration, resolved along case-fixed x- and y-axes.

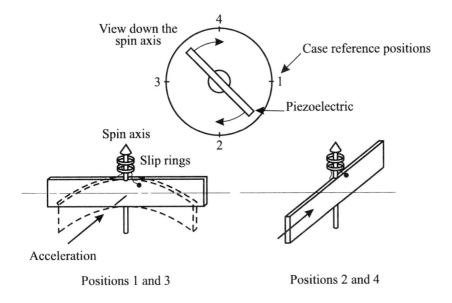

Figure 10.11. The multisensor acceleration sensor.

The gyro crystal, shown in Figure 10.12, bends in an S-shape under the gyroscopic torque produced by a rotation about the axis shown, as the crystal bends

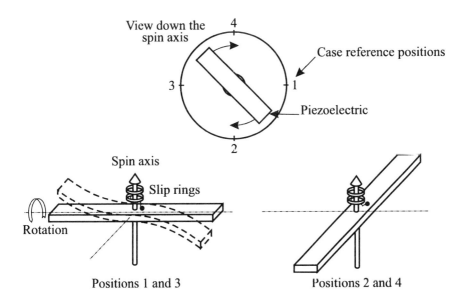

Figure 10.12. The multisensor rotation sensor.

to try to align its angular momentum with the input rate. Again, as the spin axis turns a quarter-turn, the signal falls to zero; a quarter-turn later, it repeats position 1, and so on.

The charges pass from the spinning crystals either through slip rings or a rotary transformer to the electronics, where they are filtered, amplified, demodulated, and output as x- and y-axis rates and accelerations. As multisensors are open-loop instruments, they have a poorer scale factor than SDFGs, and as the piezoelectric crystals age, the scale factor can change. However, these instruments are good enough for some missions; one of their advantages (common to all open-loop sensors) is that they can measure to high rotation rates without dissipating any torquer power. That can be a significant issue in high dynamic applications, where the SDFG and the DTG would dissipate many watts of power, which must be supplied by the servo electronics.

The British Aerospace DART is built without the accelerometer crystal, and the gyro crystals are sealed in a spherical housing filled with mercury, whose angular momentum increases the force on the crystals and increases the gyro sensitivity [18]. The Kearfott and Rockwell versions do not use the mercury and are less expensive.

There are at least three error sources that must be controlled [19]. First, any slip ring or rotary transformer noise is synchronous with the crystal signals and represents a rate or acceleration error. Second, the bearing noise is coupled to the crystals and generates an enormous background signal that can saturate the input stage of the electronics. Low-noise bearings can alleviate this problem, but they are more expensive and can be a continuing production problem. Third, the measurement bandwidth depends on the crystal spin speed; Shannon's theorem stipulates that to measure a sinusoidal signal one must have at least two measurement points per cycle. Thus a gyro of 100 Hz bandwidth must spin at least at 200 rev/s, or 12,000 rpm. If a user can live with a lower bandwidth, the spin speed can be reduced, which will also reduce the bearing noise. The best applications for this instrument (because of its small size) are in high dynamics missiles, in their seeker heads or midcourse guidance systems, both applications needing at least 100 Hz bandwidth.

This kind of gyro can have unstable bias over temperature because the bearings change their internal loading and their lubricating oil changes its viscosity, causing the noise characteristics to vary with temperature. Also, the instrument really measures a vector (acceleration or angular velocity), resolved by the electronics into x and y components. The accurate resolution of the vector needs a stable phase reference, but filters and other electronics components can change their characteristics with temperature, causing phase errors. As a result, the input axes appear to rotate over temperature, which looks to the system like a scale factor error.

Conclusion

These vibrating sensors clearly fall into two classes. One class comprises the QRS, silicon micromachined TFG, START, and DART, inexpensive and rugged sensors of accuracy 0.01–0.1 deg/s, and capable of being fired from guns. There is a four-order leap to the second class, the HRG, built with precision and art and capable of 0.001 deg/h long-term drift, at a high price. Only the gyros of the first class have seen production; the DART was used in the PaveWay 3 guided bomb, and the QRS in the Maverick missile.

References

1.   Quick, W.H., "Theory of the vibrating string as an angular motion sensor," Trans. ASME, J. Appl. Mech., pp. 523–534, Sept. 1964.
2.   Chen Feng-Yu, Quing Rong-Kang, "The development of piezoelectric crystal gyroscopes in China," DGON Symposium Gyro Technology, Stuttgart, 1988.
3.   Boocock, D., L. Maunder, "Vibration of a symmetric tuning fork," J. Mech. Eng. Sci., 11, 4, 1969.
4.   Fox, C.H.J., D.J.W. Hardie, "Vibratory gyroscopic sensors," DGON Symposium Gyro Technology, Stuttgart, 1984.
5.   Hunt, G.W., A.E.W. Hobbs, "Development of an accurate tuning-fork gyroscope," *Symposium on Gyros,* Proc. Inst. Mech. Eng. (London), 1964–65, 179, 3E.
6.   Barbour, N., J. Connelly, J. Gilmore, P. Greiff, A. Kourepenis, M. Weinberg, "Micromechanical silicon instrument and systems development at Draper Laboratory," AAIA Guidance, Navigation and Control Conference, San Diego, CA, 29–31 July 1996.
7.   Weinberg, M., J. Bernstein, S. Cho, A.T. King, A. Kourepenis, P. Maciel, "A micromachined comb-drive tuning fork rate gyroscope," Institute of Navigation, Proc. 49th Annual Meeting, Cambridge, MA, 21–23 June 1993.
8.   Lynch, D.D., "Hemispherical resonator gyro," in Ragan, R.R. (Ed.) "Inertial technology for the future," IEEE Trans. on Aerospace and Electronic Systems AES-20, 4, pp. 414–444, July 1984.
9.   Bryan, G.II., "On the beats in the vibrations of a revolving cylinder or bell," Proc. Camb. Phil. Soc., Vol VII, Nov. 24, pp. 101–111, 1890.
10.   Loper, E.J., D.D. Lynch, "Projected system performance based on recent HRG test results," Paper S83-105, IEEE/AIAA Fifth Digital Avionics Systems Conference, 31 Oct.–3 Nov. 1983.
11.   Loper, E.J., D.D. Lynch, "The HRG: A new low-noise inertial rotation sensor," 16th Joint Services Data Exchange for Inertial Systems, Los Angeles, CA, 16–18 Nov. 1982.
12.   Scott, W.B., "Delco makes low-cost gyro prototype," Aviation Week and Space Technology, 25 Oct. 1982.

13. Fox, C.H.J., "Vibrating cylinder rate gyro, theory of operation and error analysis," DGON Symposium Gyro Technology, Stuttgart, 1988.
14. Burdess, J.S., "The dynamics of a thin piezoelectric cylinder gyroscope," Proc. Inst. Mech. Engrs. (London), Vol. 200, No. C4, pp. 271–280, 1986.
15. Harris, D.G., "START: A novel gyro for weapon's guidance," DGON Symposium Gyro Technology, Stuttgart, 1988.
16. Koning, M.G., "Vibrating cylinder gyroscope and method," U.S. Patent 4 793 195, 27 Dec. 1988.
17. Pittman, R., "Rate sensor," U.S. Patent 3 359 806, 26 Dec. 1967.
18. Jones, B.G., "The development of a miniature twin-axis rate gyro," in *Mechanical Technology of Inertial Devices*, Proc. Inst. Mech. Eng. (London), 1987-2, Paper C62/87.
19. Rider, B.F., "Sensor assembly having means for cancellation of harmonic induced bias from a two-axis linear accelerometer," U.S. Patent 4 462 254, 31 July 1984.

11
The Principles of Optical Rotation Sensing

So far, we have considered inertial navigation sensors that use the mechanics of matter described by Newton's laws of motion, basically concerned with the conservation of momentum in a frame of reference fixed in the stars. In this and the following chapters we will describe gyroscopes based on the inertial property of light. We will begin by defining that property, then we will review the theory of some optics used in gyros.

The Inertial Property of Light

Maxwell's wave theory and the quantum theory can explain between them all the optical phenomena that we observe. Newton had argued that light acted as if it were corpuscular, and Huygens had proposed wave explanations for refraction and reflection. Nowadays we see both aspects of light's interaction with the rest of the physical world operating in the optical gyro.

When Maxwell published his theory of electromagnetic waves in 1860, he showed that the waves would travel through space with a speed of 3×10^8 m/s. He noted that this was the value that people had measured for the speed of light and suggested that light might be an electromagnetic wave. The *ether* was postulated as the medium that carried light; it had to permeate all space, since we can see the distant stars because the light they emit reaches us, and it had to be very rigid to allow the propagation of such high-speed waves. The ether must not have viscosity; there can be no mechanical drag associated with it, because Kepler's laws describe planetary motions without assuming any drag.

In any frame of reference (such as one fixed in the earth) moving with respect to the ether, the velocity of light would depend on the frame velocity relative to the ether, just as sound wave velocity depends on the source and observer's velocities relative to the air. So, to test for an ether, Michelson and Morley set out to measure the difference in the speed of light on earth due to the earth's orbital velocity [1]. They concluded (in 1887) that the earth did not move relative to an ether.

In his special theory of relativity (1905), Einstein postulated that the speed of light is independent of the motion of the source. Accordingly, there is no need of an ether frame of reference, and no surprise at the Michelson-Morley result.

Early in the 20th century, other scientists sought the ether by rotating a ring interferometer (one in which the light beams travel around a closed path, enclosing a finite area) in its plane; Haress (circa 1910) used a ring interferometer made of glass prisms. Not all these experimenters came up with the same answer, though; Georges Sagnac reported (1913–1914) that he saw fringe shifts that confirmed the ether's existence [2,3]. Nevertheless, the effect that relates interferometer fringe shift to rotation rate is called the Sagnac effect [4]; it is the basis of all the optical gyros we will describe in this book.

The Sagnac Effect

Sagnac's fringe shift can be explained without the need for an ether [5,6]. In the ring interferometer, shown in Figure 11.1, two light waves circulate in opposite directions around a path of radius R, beginning at source S. Assume that the source is rotating with velocity Ω, so that the light traveling in the opposite direction to the rotation returns to S sooner than that which travels in the same direction. The wave traveling with the rotation covers a distance L^+ in transit time t^+, and that against it covers L^- in time t^-. Then

$$L^- = 2\pi R - R\Omega t^-$$

$$L^+ = 2\pi R + R\Omega t^+$$

If the waves can circle the path N times, the difference in transit times is

$$\Delta t = N(t^+ - t^-) = 4\pi NR^2\Omega/c^2 = (4NA/c^2)\Omega$$

and the distance represented by this transit time difference is

$$\Delta L = c\Delta t = 4AN\Omega/c = (Ld/c)\Omega \tag{11.1}$$

where
  A = area enclosed by path
  d = path diameter
  L = total length traversed

For a wave of frequency f (wavelength $\lambda=c/f$), the period $T = 1/f$ is the time taken for a 2π phase change, so that the Sagnac phase change ϕ over time Δt is

$$\phi = 2\pi\Delta t/T = 2\pi f\Delta t$$

so that, after substituting for Δt,

$$\phi = (8\pi AN/\lambda c)\Omega \tag{11.2}$$

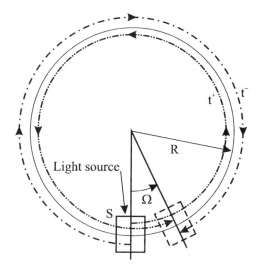

Figure 11.1. Transit time difference due to rotation.

The fringe shift x is then $\phi/2\pi$, and

$$x = (4AN/\lambda c)\Omega \ = \ (4dL/\lambda c)\Omega = K_o\Omega \tag{11.3}$$

where K_o is the open-loop scale factor. The light path can be defined by prisms (Haress), mirrors (Sagnac), or optical waveguides; it can have any shape, and A is the plan area of the shape normal to the rotation vector.

Figure 11.2 shows Sagnac's interferometer. Light from a source meets a beam splitter where it is divided into beams traveling clockwise (marked with "+" signs in the figure) and counterclockwise (marked "–") around a path defined by four mirrors. On completing the circuit the beams again meet the beam splitter and combine to pass to a camera. The two beams interfere to form a fringe pattern at the camera, which records the fringe position. The interferometer is mounted on a table that rotates it in the plane of the mirrors, and the fringe displacement is measured as a function of rotation rate. If it were to be used as a gyro, the rotation rate would be calculated from (11.2). Nowadays, the beams combining on the camera would meet on a photodetector, in which case the output current is a measure of phase shift.

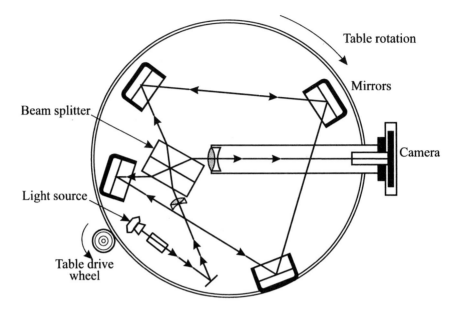

Figure 11.2. Sagnac's interferometer.

Sagnac Sensitivity—The Need for Bias

Assume a lossless Sagnac interferometer with an ideal 3-dB beam splitter. The light amplitude at the detector is the sum of the clockwise and counterclockwise amplitudes, shifted by the Sagnac phase ϕ

$$E_D = E^+ + E^- e^{i\phi}$$

Assuming $E^+ = E^- = E$,

$$E_D = E(1 + \cos \phi - i \sin \phi)$$

Then the intensity on the detector is

$$I_D = |E_D|^2 = 2I(1 + \cos \phi)$$

where $I = |E|^2$. The sensitivity is

$$|dI/d\phi| = 2I \sin \phi$$

If the interferometer is stationary in inertial space, the intensity at the detector is maximum. As the interferometer rotates faster the intensity drops, reaching zero

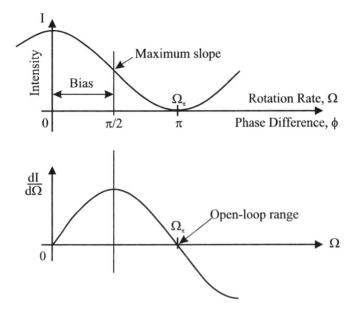

Figure 11.3. Sagnac interferometer sensitivity.

when the Sagnac phase difference is π (Figure 11.3). Also, because the slope of the detector response curve is zero at zero rate, the basic Sagnac interferometer has a high threshold; large changes from zero rate are needed to give a detectable output.

At the rotation rate where the Sagnac phase $\phi = \pi$, $\Omega_\pi = \lambda c/8AN$ [Equation (11.2)], and the intensity will again fall to zero. Once at this point, any increase in intensity can be due to either an increase or a decrease in rate; the output is ambiguous.

To overcome the first problem, the interferometer can be biased to the point of maximum slope (the maximum sensitivity) at $\phi = \pi/2$. This would give optimum threshold and reasonable linearity; the scale factor would at least be predictable, if not constant. The rate range is still limited to that which represents a $\pm\pi/2$ phase shift, where ambiguity sets in. To overcome that problem we need to close a servo on the Sagnac phase, inserting a nonreciprocal phase modulator to oppose it so that the output intensity stays at the maximum-slope bias point. The gyro rate output will then be whatever measure we have of the feedback signal to this phase modulator, usually a voltage. We will describe specific implementations in the next chapter.

The Shot Noise Fundamental Limit

When we studied mechanical gyros we did not consider them as having some theoretical limit to their performance. There is a mechanical noise threshold, Boltzmann noise, from the random thermal motion of the gyro structural material, which limits mechanical gyro performance, but this theoretical threshold is far below that which is achieved. In practice, material properties such as microcreep cause mass shifts, and electromagnetic components cause reaction torques, which overwhelm the fundamental limit.

But that is not so with optical gyros. While they have excellent stability over periods of hours, they are quite noisy in the short term (over seconds); indeed, they act as white noise generators. Their theoretical fundamental threshold noise really does set a lower limit to their performance. Mechanical gyros, especially gas bearing single-degree-of-freedom floated gyros, are quieter and preferable if ultra-quiet short-term pointing accuracy is needed, in the nanoradian regime.

Light energy is quantized in photons, and when light falls on a detector, it creates electrical noise (*photon shot noise*). Assuming that the detector amplifier electronic noise is small, the shot noise depends on the detector current and the measuring bandwidth [7], its r.m.s. value being

$$i_s = (2ei_dB)^{1/2} = (ei_d/t_s)^{1/2} \qquad (11.4)$$

where
    e = electron charge
    i_d = average detector current
    B = measurement bandwidth = $1/2t_s$
    t_s = sample time

The gyro scale factor, $K = i_d/\Omega$, is the ratio of the output signal to the input rate, so that the minimum detectable rate change, $d\Omega$, caused by the uncertainty in output current i_s, defines the resolution

$$\frac{d\Omega}{\Omega} = \frac{1}{K}\frac{di_d}{i_d} = \frac{1}{K}\frac{i_s}{i_d}$$

so that, using (11.4),

$$\frac{d\Omega}{\Omega} = \frac{1}{K}\left(\frac{e}{i_d t_s}\right)^{1/2}$$

If the rate noise has a white spectrum with an r.m.s. value σ, it gives rise to a random walk in angle, R_θ, which propagates as the square root of time

$$R_\theta = \sigma\sqrt{t} \qquad (11.5)$$

Angle random walk is a commonly specified parameter for optical gyros.

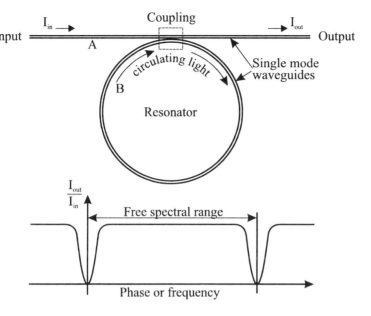

Figure 11.4. The optical resonator.

The Optical Resonator

In the Sagnac interferometer, a light wave splits and passes both ways around a path before being recombined and interfering with itself. The beam travels once over the path, making this a *single-beam* interferometer. The Mach-Zehnder (Figure 3.10) is another kind of single-beam interferometer; as with the Sagnac, its output varies sinusoidally (actually a raised cosine) with phase. The light frequency is constant in a single-beam interferometer.

Resonators can operate with different clockwise (CW) and counterclockwise (CCW) frequencies. If we have a path in which light can travel (it could be a ring of single-mode optical fiber) with some means of coupling light in and out, as shown in Figure 11.4, and if the perimeter of the ring is an integral number of wavelengths of the light in the medium, the ring will resonate. If this resonator rotates, the Sagnac phase shift causes the CW and CCW resonances to change. (Compare this with the serrodyne frequency shifter, discussed in Chapter 12.) For resonance the phase condition at a point must be the same for the wave leaving the point as it is for the wave arriving at the point, so that any Sagnac phase shift must be compensated by a shift in the wavelength to preserve phase matching.

Light enters the resonator of Figure 11.4 through the coupler and circulates. When it gets to the coupler again, its field is added to that of the wave entering through the coupler; the resulting amplitude depends on the phase between them. The wave circulates around the same path many times, so this kind of interferometer is called a *multiple-beam* interferometer. The continuing interference

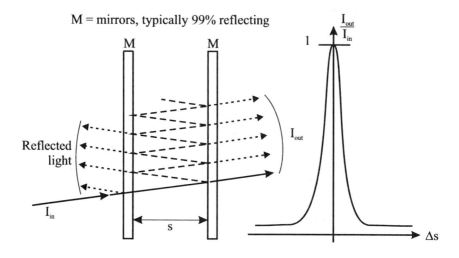

Figure 11.5. The Fabry-Perot interferometer.

in the resonator sharpens its frequency response, so that it is narrower than the sinusoidal response of the single-beam interferometer. Optical losses such as absorption and scattering broaden the linewidth, because the waves can only circulate a few times before their energy is dissipated to the level of the background. If the loss is so high that the wave only makes it around the ring once, the frequency response is that of the single-beam interferometer—sinusoidal.

The Fabry-Perot Resonator

The classical Fabry-Perot interferometer is a free-space optics device made from a pair of parallel, high-reflectance mirrors, shown in Figure 11.5. It is sometimes called an *etalon* [8]. For the sake of example, assume that the mirrors have 99% reflectivity and are lossless. Light enters through the left mirror (1% transmission), passes across distance s to the right mirror, where 99% is reflected and 1% transmitted. After reflection at the left mirror, another 1% passes out of the right mirror, interfering with the first. This continues with subsequent reflected and transmitted waves. The phases of the transmitted waves, and thus the net intensity of the output, depend on the mirror spacing, s. When the roundtrip phase is an integer multiple (m) of 2π, the etalon resonates and the output rays are precisely in phase with one another. There will be no light reflected, as the reflected rays will be in antiphase, and, in a lossless system, the output intensity will be equal to the input intensity—the transmissibility will be 1. There will be a resonance for each integer value of m.

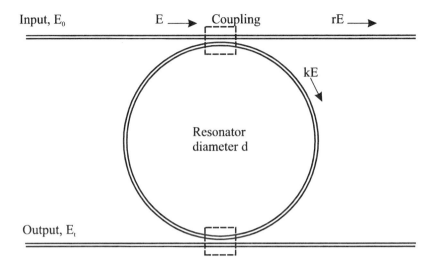

Figure 11.6. A two-port waveguide resonator.

The same process operates in the waveguide resonator, shown in Figure 11.6 with two couplers analogous to the two mirrors of the etalon. It has diameter d and perimeter $p = \pi d$. In order to derive the expression for the power transmission of such a resonator, assume that the waveguide has a field attenuation of α per unit length, so that the field along the guide is

$$E(z) = Ee^{-\alpha z}$$

Assume also that the couplers are lossless and have a field transmission coefficient k. Then if a wave of amplitude E_o enters the input coupler, the amplitude just inside the resonator is kE_o. If the output coupler is opposite the input coupler, the wave amplitude at the output coupler is $kE_o e^{-\alpha p/2}$.

The amplitude of the wave transmitted through the output coupler is

$$\Delta E_{t1} = k^2 E_o e^{-\alpha p/2} \tag{11.6}$$

The amplitude in the resonator just past the output coupler is $rkE_o e^{-\alpha p/2}$, where r is the coupler reflection coefficient. Going around the ring to the input coupler, the amplitudes just before it and just after it are $rkE_o e^{-\alpha p}$ and $r^2 kE_o e^{-\alpha p}$. Back at the output coupler, the amplitude to be added to (11.6) is

$$\Delta E_{t2} = r^2 k^2 E_o e^{-\alpha p} \tag{11.7}$$

and so on for an infinite number of trips around the resonator. Guided waves are described by the *propagation constant* $\beta = 2\pi n/\lambda_0$, where n is the effective index and λ_0 is the free space wavelength; waves having different propagation constants travel at different speeds. If we assume that there is no phase change in the couplers, the phase difference δ between the waves at the output coupler is $\delta = \beta p$.

Summing the transmitted waves whose first terms are (11.6) and (11.7), allowing for the phase $e^{i\delta}$,

$$E_t = k^2 E_o e^{(i\delta - \alpha p)/2}[1 + r^2 e^{(i\delta - \alpha p)} + r^4 e^{2(i\delta - \alpha p)} + \cdots]$$

Summing to infinity

$$E_t = \frac{k^2 E_o e^{(i\delta - \alpha p)/2}}{1 - r^2 e^{(i\delta - \alpha p)}}$$

The power transmitted is the square of this complex number

$$\frac{P_t}{P_o} = \frac{k^4 e^{-\alpha p}}{[1 - r^2 e^{(i\delta - \alpha p)}]^2}$$

If we let $a = r^2 e^{-\alpha p}$, the denominator simplifies to

$$(1-a)^2 + 4a \sin^2 \delta/2$$

The power transmission and reflection coefficients are $R = r^2$ and $K = k^2$, and for lossless couplers $R = 1 - K$, so that

$$a = (1-K)e^{-\alpha p}$$

$$\frac{P_t}{P_o} = \frac{K^2 e^{-\alpha p}}{[1 - (1-K)e^{-\alpha p}]^2} \cdot \frac{1}{1 + A^2 \sin^2 \delta/2} \tag{11.8}$$

$$A^2 = \frac{4(1-K)e^{-\alpha p}}{[1-(1-K)e^{-\alpha p}]^2}$$

Thus, Equation (11.8) gives the phase (or frequency) response of a Fabry-Perot resonator in terms of its waveguide loss and coupling constant.

The peak transmitted power occurs at $\delta = 0, 2\pi, ..., 2m\pi$, where m is an integer, the longitudinal mode number. The peak power ratio (P_t/P_o) is given by

$$P = \frac{K^2 e^{-\alpha p}}{[1-(1-K)e^{-\alpha p}]^2} \tag{11.9}$$

For a lossless guide $(\alpha = 0)$, $P = 1$. As mentioned earlier, the Fabry-Perot etalon has a series of resonances, for each integer m, whose frequency spacing is the *free spectral range* (FSR). For a ring of perimeter p, the FSR is

$f_{sr} = c/np$

Resonator Finesse

The quality factor, Q, was defined in Chapter 3 for the frequency response of a mechanical resonator. The same definition holds for an optical resonator

$$Q = f_o/\Gamma$$

where
   f_o = resonant frequency
   Γ = resonance linewidth (full width at half maximum)

But whereas the Q-factor is a useful parameter when the resonant frequency is in the kHz or MHz range (10^2 to 10^8 Hz), optical frequencies are so high (10^{14} Hz) that optical Qs would be meaninglessly large. In interferometry, instead of using the absolute resonant frequency as a reference, we use the free spectral range, defining a parameter analogous to the Q-factor called the *finesse*, F:

$$F = f_{sr}/\Gamma = c/np\Gamma \tag{11.10}$$

From Equation (11.8),

$$F = \frac{\pi A}{2} = \frac{\pi(1-K)^{1/2}e^{-\alpha p/2}}{1 - (1-K)e^{-\alpha p}} \tag{11.11}$$

For a lossless guide, $F = \pi(1-K)^{1/2}/K$. The finesse represents the number of times a wave circulates around the resonator until its intensity has dropped to the background light level; it is analogous to the number of turns in a fiber Sagnac interferometer of the same dimensions [9].

The Sagnac Effect in a Resonator

If we assume a resonator operating with a longitudinal mode number m so that $L = m\lambda = mc/f$, then $dL/df = -L/f$. The effect of the Sagnac length change can be deduced from Equation (11.1):

$$\Delta L = 4AL\Omega/cp; \quad |\Delta L/L| = \Delta f/f = 4A\Omega/cp$$

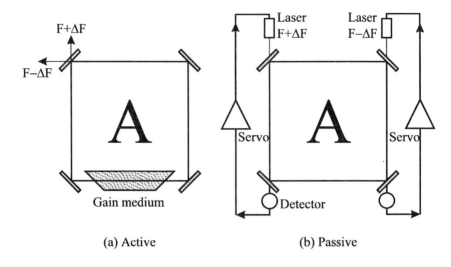

Figure 11.7. Active and passive resonators.

In a medium of index n, c/n = fλ, so

$$\Delta f = \frac{4A}{n\lambda p} \Omega \tag{11.12}$$

If we are using this frequency change on rotation to make a gyro, Equation (11.12) defines the scale factor $K = \Delta f/\Omega$

$$K = 4A/(n\lambda p) \tag{11.13}$$

For a ring resonator, $K = d/n\lambda$. (We will find in Chapter 12, Equation (12.1), that this is the same as the scale factor for the phase-nulled interferometric fiber-optic gyro.)

If the frequency is thought of in pulses/s, then the reciprocal of K represents a pulse weight in arc-sec/pulse, or if each pulse is a digital bit, in arc-sec/bit.

The number of circuits of the path, N, does not affect the scale factor, but does affect the theoretical gyro threshold, or minimum detectable rate, by way of the resonator finesse.

Active and Passive Resonators

The resonators shown in Figures 11.4–11.6 are optical circuits with light entering either through a coupler or through a mirror. It is possible to place a lasing medium such as an excited helium-neon gas mixture inside a free-space resonator, so that light is generated inside the resonator and automatically finds its CW and CCW resonant frequencies. We call this an *active* resonator; see Figure 11.7a.

Alternatively, we can probe the resonator with laser beams from outside, by placing photodetectors so that they measure the intensities of the CW and CCW beams, as shown in Figure 11.7b. We can servo the frequencies of the light sources so that the resonator remains on resonance in both directions; this configuration is called a *passive* resonator. The active resonator emits the CW and CCW resonant frequencies, whereas in the passive resonator we must find them. Active and passive resonator gyros have identical scale factor and fundamental limits for identical parameters, but the achieved resolution of these gyros depends on necessary engineering choices that prevent the ideal from being achieved.

Resonator Figure of Merit

Equation (11.11) expresses the finesse of the resonator of Figure 11.6, with coupling K and distributed waveguide loss of α cm^{-1}.

$$F = \frac{\pi(1-K)^{1/2}e^{-\alpha p/2}}{1 - (1-K)e^{-\alpha p}}$$

Note that finesse falls monotonically with increasing coupling, because the power coupled out looks like loss to the resonator. The coupling can be reduced to get high finesse, but only at the expense of the power transmitted to the detector. The resonance may then be very sharp, but it cannot be seen (it becomes unobservable). We can define a parameter that combines finesse and observability into a figure of merit

$$F_m = F(P_t/P_o)^{1/2} \tag{11.14}$$

where P_t/P_o is the transmitted power [Equation (11.8)].

If the resonator is overcoupled, the transmitted power is high and the finesse will be reduced. There is an optimum coupling corresponding to the peak F_m; it occurs when the coupling K is equal to the roundtrip loss, αp, which we will use when we consider the fundamental noise performance of the passive resonator.

Optical Fibers

The Sagnac interferometer of Figure 11.2 has low sensitivity because the light passes only once around the path [N = 1 in Equation (11.2)]. One way to make a Sagnac interferometer more sensitive is to send the light around the path many times (N>1), by replacing the mirrors with an optical fiber coil. Although mirrors and free-space optics preceded fiber guided-wave optics by centuries, we will leave

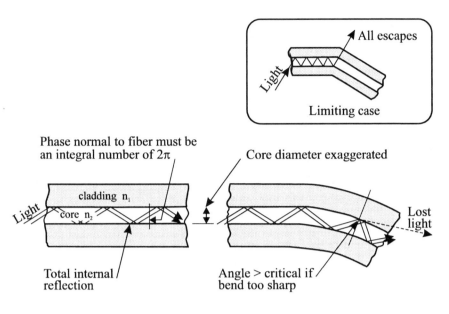

Figure 11.8. Transmission in a fiber.

the discussion of free-space optics gyros until later and first describe some of the fundamentals of the optical fiber.

The optical fiber guides light by total internal reflection, a consequence of refraction and the critical angle. The next section reviews these characteristics of light propagation in a medium.

Refraction and Critical Angle

The refractive index of a transparent medium, n, relates the speed of light in the medium (v) to that in free space (c) by $v = c/n$; the index is usually temperature dependent. The differing speed in different media causes refraction; Snell's law relates the refraction of a wave to the refractive index of the medium

$$n_1 \sin \theta_1 = n_2 \sin \theta_2$$

where

n_1, n_2 = media refractive indices

θ_1, θ_2 = angles of incidence (to the normal)

If $n_1 > n_2$, there will be a value of θ_1 for $\theta_2 = 90°$, called the *critical angle*. If θ_1 is increased further, no light passes into medium 2, because it is totally internally reflected within medium 1. This property underlies the operation of the optical fiber.

Multimode and Single-Mode Fibers

An optical fiber is a fine thread of glass composed of a core of higher index material in a cladding material, commonly silica. The core diameter is of the order of the wavelength of light. Light passing along it is incident on the boundary between the core and the cladding at an angle greater than the critical angle, so that it reflects along the core as illustrated in Figure 11.8. However, not any ray incident at greater than the critical angle can propagate, since it is necessary that the phases of the reflected rays reinforce one another, so that the phase differences must be multiples of 2π, illustrated in the figure. Consequently there are incidence angles that allow rays to propagate, and each is a different mode. If the fiber size and index are chosen appropriately, there is only one mode allowed at a particular wavelength [10]. Such a fiber is called a *single-mode* fiber in the United States and a *monomode* fiber in Europe. *Multimode* fibers have core diameters much larger than the light wavelength.

Light escapes at a bend if the curvature is too large and/or there is only a small difference between the core and cladding indices. This is because the incident ray no longer meets the outside of the bend at greater than the critical angle, as shown in Figure 11.9. This effect causes light loss at microbends and kinks in the fiber.

Fibers are drawn from a composite glass preform. A preform is about 5 cm in diameter and 2 m long and is held vertically at the top of a tower some 10 m tall. The preform's lower tip is melted in an annular furnace and a fine thread drawn down; the thread has the same cross-sectional composition as the preform on a microscopic scale. The drawing speed determines the fiber diameter. Typically, the fiber core is a few micrometers in diameter, and the cladding is of the order of 100 µm in diameter. The fiber is usually polymer coated for protection so that a complete fiber has a diameter of 250 µm.

The fiber behaves as if its core's refractive index is slightly different from that of the bulk material, and we can characterize the fiber as having an effective index, which may be dependent on both temperature and light wavelength. The effective length of an optical fiber is its actual length multiplied by its effective index.

Polarization

The *polarization* of an electromagnetic wave is defined as the direction in which the electric field propagates. A fiber drawn from a cylindrical preform has a circular cross section, so it is degenerate in polarization; there is no preferred axis for aligning polarization. But refractive index generally depends on the stress in the

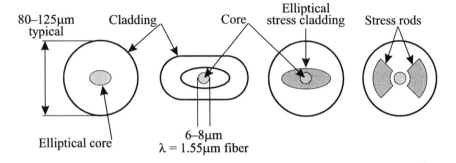

Figure 11.9. Some polarization maintaining fibers.

material, because stress causes the core to act as if it had different refractive indices in different radial directions, called *stress birefringence*. Therefore an actual fiber coil may provide a preferred polarization, although it may not be stable over temperature and time.

A birefringent fiber can be considered as having different propagation constants β_x, β_y along two orthogonal axes x and y normal to the fiber axis, so that the waves traveling along x and y travel at different speeds, and linearly polarized light coupled into both x and y will become elliptically polarized as it moves along the fiber. When it has traveled a certain distance, the *beat length*, L, the phase between the x and y wave components will have changed by 2π and the wave will again be linearly polarized [11]

$$|\beta_x - \beta_y| L = 2\pi$$

Another measure of fiber birefringence is the amount of power transferred between the cross-polarized states per unit length [12], denoted by the h-parameter. If linearly polarized light is launched into the x-axis of 1 km of a fiber with $h = 10^{-4} \, \text{m}^{-1}$, 10% of the power will be in the y-axis mode at the other end.

Birefringent Fiber for a Sagnac Gyro

The accurate measurement (or nulling) of the small rotation-induced Sagnac phase requires the rejection of the large phase (e.g., $10^6\pi$) common to the CW and CCW paths. There are two ways of ensuring this; either scramble the polarization so that it is random at the detector [12] or ensure that the paths have identical optical lengths (identical βs). Practically, that implies that they use the same polarization. Most gyro developers seem to be using the latter approach, employing special polarization maintaining fibers. Figure 11.9 illustrates some of the ways that these are made. The direct way is to make the core elliptical [13], although circular-section cores give the lowest losses. It is better to induce stress birefringence in the core by providing orthogonal asymmetry; successful methods include:

1. building-in an elliptical member of a different expansion glass,
2. building-in two rods (*stress rods*) on either side of the core—the so-called "Bow-Tie" fiber [14] and the PANDA fiber [15], for example—and
3. flattening the preform by squeezing it while it is hot [16], the so-called "square" fiber.

Square fiber has the useful property that the asymmetry axis is clearly visible from outside and coils with controlled core position can be readily wound. Stress rods function by setting up an asymmetric stress in the core when the fiber cools down.

The Coherence of an Oscillator

We will find that the performance of optical gyros depends on the characteristics of the light source. In the case of the fiber-optic gyro we are particularly concerned with source coherence.

The wave generated by an oscillator is said to be coherent if its phase is continuous. A source can have degrees of coherence depending on the length of time over which the phase is continuous (or the distance the wave propagates with an unbroken wave form). The coherence length of the emission from a light source with a linewidth Γ (the full width at the half power point) is $L_c = c/\Gamma$; the narrower the line, the longer the coherence length.

The domestic electricity supply provides a coherent 60 Hz (in the United States) voltage, whereas the crackling heard in a radio receiver in a thunderstorm comes from the incoherent RF energy emitted by a lightning flash. The laser is a coherent source of light, electromagnetic energy at about 10^{14} Hz, whereas the light from an electric light bulb is incoherent, coming as it does from the random emission of photons from the heated filament.

Types of Optical Gyro

Optical gyroscopes may be characterized as either passive or active and as resonant or nonresonant. In passive sensors the Sagnac phase is measured by some external means, whereas in active gyros the Sagnac phase causes a frequency change internal to the gyro that signals the rotation directly.

The Sagnac interferometer is the basis of the Interferometric Fiber-Optic Gyro (IFOG), which is becoming more widely used in tactical and aided navigation. The most widely used optical gyro is the active resonant ring laser gyro (RLG), a very successful product that has dominated the 1-m/h aircraft navigator business since 1980, although it may well be supplanted by the IFOG in the early years of the next century. Two resonant passive gyros, the Resonant Fiber Optic Gyro (RFOG) and the Micro-Optic Gyro (MOG), are potentially lower-cost instruments than the RLG. They showed promise in the 1980s but have languished in the 1990s. We will study the RLG in Chapter 13, and the two passive resonant gyros in Chapter 14.

Conclusion

Optical gyros have different operating characteristics from mechanical gyros. Instead of the mechanical noise created by ball bearings and the electrical noise from pickoffs, the optical gyros' noise floor is set by quantum effects (shot noise and spontaneous emission). The optical gyro's threshold is relatively high, compared with the stiction in the mechanical gyro, but the optical gyro's long-term stability is an order of magnitude better. Rosenthal [17] and Anderson [18] provide good overviews of optical gyro technology.

References

1. Tipler, P.A., *Foundations of Modern Physics*, Worth Publishers, Inc., New York, Chapter 1, 1973.
2. Sagnac, G., "L'éther lumineux demontré par l'effet du vent relatif d'éther dans un interféromètre en rotation uniforme," C. R. Acad. Sci., 95, pp. 708–710, 1913.
3. Sagnac, G., "Sur la preuve de la réalité de l'éther lumineux par l'expérience de l'interférographe tournant," C. R. Acad. Sci., 95, pp. 1410–1413, 1913.
4. Hariharan, P. "Sagnac or Michelson-Sagnac interferometer?" Appl. Opt., 14, 10, pp. 2319–2321, Oct. 1975.
5. Post, E.J., "Sagnac effect," Rev. Mod. Phys., 39, 2, pp. 475–493, April 1967.
6. Arditty, H.J., H.C. Lefèvre, "Theoretical basis of Sagnac effect in fiber gyroscopes," in Ezekiel, S., H.J. Arditty (Eds.), *Fiber-Optic Rotation Sensors*, Springer-Verlag, New York, p. 44, 1982.
7. Yariv, A., *Introduction to Optical Electronics*, Holt, Rinehart, and Winston, Inc., New York, Chapter 7, 1971.
8. Fowles, G.R, *Introduction to Modern Optics*, Holt, Rinehart and Winston, Inc., New York, 1968. Fabry-Perot interferometer, p. 86 et seq.
9. Lawrence, A.W., "The Micro-Optic gyro," DGON Symposium Gyro Technology, Stuttgart, 1983.
10. Marcuse, D., *Theory of Dielectric Optical Waveguides*, Academic Press, New York, 1974. Chapter 1 uses a geometric optics approach.
11. Kaminow, I.P., "Polarization maintaining fibers," in Ezekiel, S., H.J. Arditty (Eds.) *Fiber-Optic Rotation Sensors*, Springer-Verlag, New York, pp. 169–177, 1982.
12. Ulrich, R., "Polarization and depolarization in the fiber-optic gyroscope," in Ezekiel, S., H.J. Arditty (Eds.) *Fiber-Optic Rotation Sensors*, Springer-Verlag, New York, pp. 52-77, 1982.

13. Dyott, R.B., "Elliptically cored polarization holding fiber," in Ezekiel, S., H.J. Arditty (Eds.) *Fiber-Optic Rotation Sensors*, Springer-Verlag, New York, pp. 178–184, 1982.
14. Birch, R.D., D.N. Payne, M.P. Varnham, E.J. Tarbox, "Fabrication and development of polarisation-maintaining fibres using gas phase etching," Proc. 1st International Conference on Optical Fibre Sensors, IEE, London, pp. 83–86, April 1983.
15. Himeno, K., Y. Kikuchi, N. Kawakamu, O. Fukuda, K. Inada, "A high extinction ratio and low loss single mode single polarization optical fiber," *Optical Fiber Sensors*, 1988 Technical Digest Series, Vol. 2, Optical Society of America, Washington, D.C., 1988.
16. Stolen, R.H., W. Pleibel, J.R. Simpson, "High-birefringence optical fibers by preform deformation," J. Lightwave Tech., LT-2, 5, pp. 639–641, 1984.
17. Rosenthal, A.H., "Regenerative circulatory multiple-beam interferometry for the study of light propagation effects," J. Opt. Soc. Am., 52, 10, pp. 1143–1148, Oct. 1962.
18. Anderson, D.Z., "Optical gyroscopes," Scientific American, 254, 4, pp. 94–99, April 1986.

12
The Interferometric
Fiber-Optic Gyro

In this chapter we show how the basic Interferometric Fiber-Optic Gyro (IFOG), the Sagnac interferometer described in Chapter 11, is operated at its maximum sensitivity point, closed-loop. We will describe phase nulling and the serrodyne feedback system, and then we will consider some of the IFOG's error sources. We will also mention some of the factors that affect the cost of the IFOG.

The History of the Fiber-Optic Gyro

Equation (11.2) expresses the phase shift developed between the clockwise (CW) and counterclockwise (CCW) circulating waves of the Sagnac interferometer when it is rotated in its plane. In the interferometer made with mirrors (Figure 11.2) the light makes only one trip around the path before hitting the detector, resulting in too low a sensitivity for a useful gyro. But in 1968, R.B. Brown of the Naval Research Laboratories suggested using more circuits to increase the sensitivity, employing a coil of optical fiber as the sensing path. However, at that time fiber losses were too large for this to be practical.

By 1975 low-loss single-mode fiber could be obtained, but it was not certain that light could be coherently transmitted through the fiber in both directions. Perhaps the phase fronts would be scrambled so that the waves emerging from the fiber would not form an interference fringe pattern. This concern was put to rest in 1975 when Vali and Shorthill (University of Utah) demonstrated fringes with light transmitted both ways around a single-mode fiber [1, 2].

Development of the IFOG proceeded in the United States (led by Ezekiel at MIT and Shaw at Stanford), in France, and in Germany. By 1980 it was clear that navigation-grade IFOGs were not easily made; reported performance had reached a plateau well below the early predicted level. Problems had been identified with the fiber, the light source, and the architecture of the gyro. At a conference at MIT in November 1981, 45 papers expounded the state of the IFOG art [3]; by 1989 some 570 papers had been published worldwide. Smith [4] has collected 107 of these papers into one volume, an invaluable asset for those wishing to study IFOGs in greater depth.

In the following sections we will look at IFOG's problems and summarize the state of the art in 1998. But first let us calculate the magnitude of this Sagnac effect we are using to measure rate. Equation (11.1) provided a relationship between the distance represented by the Sagnac phase due to rotation:

$$\Delta L = (Ld/c)\Omega$$

where
L = fiber length
d = coil diameter

Assume that d = 10 cm and L = 1 km. For two values of rate, the absolute and relative length changes are:

Rate, Ω	ΔL, cm	$\Delta L/L$
1 deg/h	10^{-11}	10^{-13}
1 deg/s	5×10^{-6}	5×10^{-10}

To imagine the size of these ΔL values, recall that the diameter of a hydrogen atom is about 10^{-8} cm. For measurement to this precision, the CW and CCW paths must have near-perfect common mode rejection— they must be absolutely reciprocal.

The Basic Open-Loop IFOG

Let us begin with a basic IFOG, as shown in Figure 12.1, the fiber analog to Sagnac's apparatus (Figure 11.2). Light from the source passes into a 50% coupler, where it is split into two waves that pass around the fiber coil in opposite directions. When the light returns to the coupler the waves combine and split, half going back to the source and half to the detector. The intensity of the light at the detector is a measure of the interference between the beams, and thus of the rotation rate of the coil in its plane. Because the intensity can also change from other causes such as variation in light source output (say, with temperature) and instability in the coupler, suitable compensation for these effects has to be made.

In Chapter 11 we described how the Sagnac interferometer could be biased to its maximum sensitivity point and used as an open-loop gyro, although it would have a limited range due to the sensitivity falling to zero at some rotation rate. Is this really a problem? To calculate the range, assume a coil of area 100 cm² with N = 1000 turns, operating at a wavelength of 1.3 μm. The open-loop zero point ($\phi = \pi/2$) occurs [Equation (11.2)] at a rotation rate Ω:

$$\Omega = \lambda c/16AN = 150 \text{ deg/s approximately.}$$

This is not high enough; most vehicles can experience short-term rates to 250 deg/s, and many go to 500 deg/s. One solution would be to reduce the coil area A

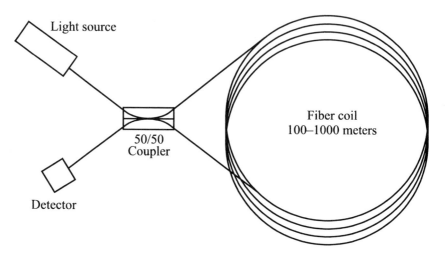

Figure 12.1. The basic interferometric FOG.

and/or the number of turns, N, but we will see later that we need to have a high AN for low resolution. Therefore the open-loop IFOG can only be used either in a low-accuracy inertial navigation system or in a heading and attitude reference system where the rates are low.

Biasing the IFOG

To bias the IFOG to the maximum sensitivity point, we need an optical element that will make the path length appear longer for one circulating wave than for the oppositely traveling wave. For $\pi/2$ bias we need a path nonreciprocity of a quarter of the wavelength of the light in the fiber.

Common phase modulators using the electro-optic effect are reciprocal, that is, the phase shift is the same regardless of the direction of the light through them. So if the CW and CCW interferometer waves pass through at the same time they will both be shifted and we will get no bias.

Nonreciprocal Phase Shifting

Nonreciprocity can be created by passing the CW and CCW waves through a time-varying phase change at different times. As light travels in fiber at 2×10^8 m/s,(n = 1.5), it travels through 1 km of fiber in 5 µs. In principle we can get a non-reciprocal effect by putting a reciprocal phase shifter at one end of the fiber coil and driving it so that the phase varies by the necessary bias with a period of 5 µs.

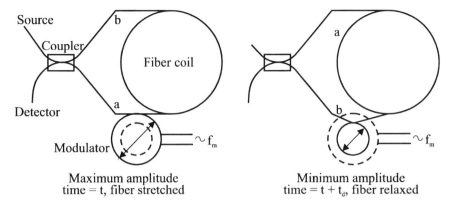

Maximum amplitude
time = t, fiber stretched

Minimum amplitude
time = t + t_d, fiber relaxed

Figure 12.2. Piezoelectric biasing device.

In one approach, some of the fiber near the input coupler was wrapped around a piezoelectric cylinder. The cylinder was driven by an alternating voltage so that it alternately stretched and relaxed the fiber [5]. To visualize the biasing process, imagine a "packet" of light entering the input coupler and dividing to go by the two available routes around the fiber, as in Figure 12.2. One part of the packet, a, immediately meets the phase modulator, which we will assume is at maximum drive and fully stretched, so the packet-part has to travel the extra length. After the coil transit time, it gets back to the coupler.

The second part of the packet, b, travels for the same time and then meets the modulator. By now the alternating drive voltage has reached the other extreme, so that the fiber is no longer stretched. So when a and b recombine into an output packet at the coupler, they have a phase difference. The phase shift $\Delta\phi$ is approximately given by [6]

$$\Delta\phi = 2\phi_0 \sin \pi f_m t_d$$

where
ϕ_0 = amplitude of phase modulation
f_m = modulation frequency
t_d = transit time around the fiber = nL/c

The phase shift is a maximum when $f_m = 1/2t_d$; for 1 km of fiber the optimum f_m = 100 kHz. Lefèvre et al. [7] give an exact analysis of this approach and show that the optimum f_m is somewhat different from the approximate value given earlier.

As $f_m = c/2nL$, and both the fiber length (L) and the effective index (n) change with temperature, the frequency must be appropriately controlled.

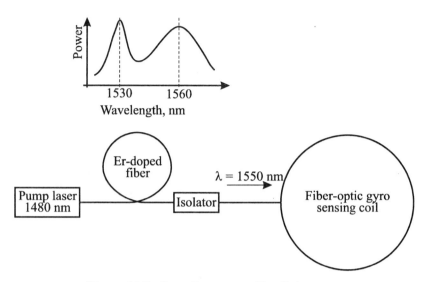

Figure 12.3. Superfluorescent fiber light source.

The Light Source

The first IFOG researchers used lasers as the source because lasers are highly coherent (Chapter 11) and give very visible fringes. However, it was quickly discovered that the light reflected at the couplers and the joins inside the fiber loop and the light Rayleigh scattered by the fiber molecules merged with the true rotation-induced signal and swamped it [8,9]. In the basic Sagnac interferometer there is only one optical frequency present, and coherent backscattered light cannot be distinguished from direct light.

The Sagnac interferometer has two interfering beams that travel almost identical path lengths, therefore it is able to form fringes with white (low coherence) light. By using a less coherent source (closer to white light), the backscattered light is incoherent and will not generally form fringes with the direct light. Low-coherence light sources have a wide linewidth whose effective wavelength (the *interferometric wavelength*) may not be well known. We saw in Equation 11.3 that the gyro scale factor depends inversely on wavelength, so it must be known and stable, to parts per million, for a good-quality navigator.

Early IFOGs used the superluminescent diode (SLD) and the edge-emitting light emitting diode (ELED) as sources. Early SLDs provided output power less than 1 mW, although recently developed models give 10–20 mW. But SLDs have wavelength/temperature coefficients of 400 ppm/°C, have an astigmatic output that is not easy to couple to a fiber, and are expensive. (They cost $5000–$10,000 each in small quantities because they are not production devices and only fiber sensor manufacturers need them.)

Semiconductor lasers are mass-produced for consumer products like CD players, so they are very inexpensive. An alternative low-cost approach to destroying coherence is to chirp the frequency of a semiconductor laser by modulating its current [10, 11]. However, it still has the same thermal sensitivity. Therefore, semiconductor sources need to be temperature controlled to better than 0.001 °C for navigation-grade gyro performance.

Taking advantage of work in the telecommunications field on fiber amplifiers, IFOG sources are now commonly made from Erbium-doped fiber [12]. Such a source is called the *Superfluorescent Fiber Source*, SFS. Output power can exceed 50 mW, linewidth 50 nm, and wavelength/temperature coefficient is less than 5 ppm/°C. They also operate at a wavelength around 1.55 μm, which is matched to the lowest-loss telecommunications fibers. However, they are very large, orders of magnitude larger than the SLD, and they are orders of magnitude more sensitive to nuclear radiation. Nevertheless, they can be readily packaged in a gyro coil, and so they have found widespread use.

Figure 12.3 shows one simple form of the SFS in which a piece of Er-doped fiber 10–20 m long is excited by the light from a 1480-nm laser passing along its core. The excited Er atoms spontaneously decay, emitting light in two peaks at 1530 and 1560 nm. By choosing fiber length and pumping configuration these two peaks can have similar outputs, providing the required wide line source. An optical isolator serves two purposes: it prevents the pump light from passing into the sensing coil; and it stops light reflected from the rest of the gyro, which could cause the fiber to lase (thus narrowing the spectral line). Particular designs to accomplish this are currently proprietary.

Reciprocity and the "Minimum Configuration"

The simplest IFOG (Figure 12.1) is not reciprocal, that is, the CW and CCW light paths have not covered identical paths when they are recombined. The light from the source is reflected in the coupler, passes CW around the coil, is reflected again at the coupler, and passes to the detector. It has two reflections in the coupler. The other beam passes through the coupler, circulates CCW, again passes through the coupler and then to the detector. It has two transmissions through the coupler. It is probable that the phase conditions for reflection and transmission at the coupler are different and do not track with temperature.

Shaw and his co-workers [13] at Stanford built an all-fiber (open-loop) IFOG, which they made fully reciprocal by adding another coupler, as shown in Figure 12.4; each beam undergoes the same number of reflections and transmissions. This layout is called the *minimum configuration*. They used polarization maintaining fiber and polarization controllers to improve the gyro performance and reported a steady-state "tombstone" drift of 0.2 deg/h with a sample time of 30 s. They made all the components in a string along the fiber without breaking it, to reduce interface reflections.

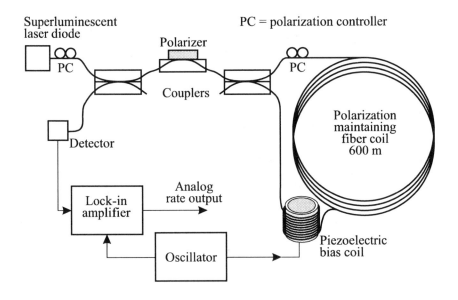

Figure 12.4. Stanford all-fiber FOG.

Closing the Loop—Phase-Nulling

While making all the components in one fiber string reduced reflections, one could not make a closed-loop gyro this way because there was no effective fiber modulator available. Because of this difficulty, the industry has not followed the all-fiber approach. One approach to phase-nulling, used by Cahill and Udd in their IFOG [14], creates a phase shift from a frequency shift.

To illustrate this, Figure 12.5 shows CW and CCW wave trains, each containing m periods, the CCW wave in length L and the CW in L+ΔL. ΔL represents the path difference due to the Sagnac phase from rotation. The consequent fringe shift for a coil of N turns is given by Equation (11.3):

$$x = 4AN\Omega/\lambda c$$

The fringe shift equivalent to the difference Δf between the CW and CCW waves is $\Delta f t_d$. Then,

$$\Delta f t_d = \Delta f n L/c = (4AN/\lambda c)\Omega, \text{ or } \Delta f = (d/n\lambda)\Omega$$

Thus the Sagnac path difference can be nulled by changing the frequency of the light circulating in one direction around the coil. Davis and Ezekiel [6] used acousto-optic Bragg frequency shifters in their phase-nulled gyro.

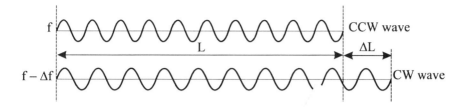

Figure 12.5. Frequency shift phase nulling.

The phase-nulled gyro's scale factor is

$$K = d/n\lambda \tag{12.1}$$

Acousto-Optic Frequency Shifters

The Bragg cell is a device that changes the frequency of a light wave. In the discrete component (free-space optics) Bragg cell, a beam of light passes through a transparent block through which a sound wave is traveling, the wave crossing the light's path. The sound wave (typically 40–80 MHz) creates a moving density pattern in the medium, which creates a moving refractive index pattern. This acts like a diffraction grating moving across the light; it diffracts the light and Doppler shifts its frequency. The light must be incident at a particular angle to the path of the sound wave, the Bragg angle, and the shifted light is also diffracted at the Bragg angle. Therefore the shifted light is spatially separated from any unshifted light by twice the Bragg angle, which can be useful. However, Bragg shifters consume 100–500 mW of RF power, which is high for gyros.

The discrete component Bragg cell is difficult to integrate into an otherwise all-solid-state gyro, so thin-film, guided-wave versions have been developed in integrated optics. While we will describe integrated optic Bragg shifters in Chapter 14 (they are more useful for resonant gyros than interferometer gyros), we will next introduce the field of integrated optics, as it has important applications to the IFOG.

Integrated Optics

Free-space optics uses lenses, mirrors, and modulators made with bulk crystals. Guided-wave optics uses dielectric waveguides to conduct light waves and interactions with electric and acoustic fields to modulate the light. The most common optical waveguide is the optical fiber (Chapter 11), but we can also make waveguides by changing the refractive index inside a glass substrate, perhaps by

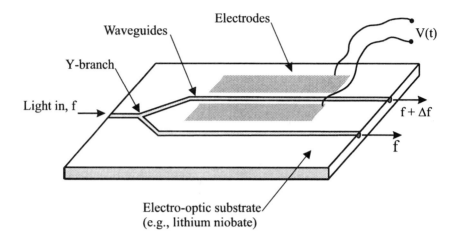

Figure 12.6. An optical chip modulator.

ion implantation, by ion exchange, or by other methods. There is a good general description of the field by Yariv [15].

To modulate light we use an active electro-optic crystal; lithium niobate is most often used. Waveguides are made in lithium niobate by diffusing titanium into the surface of the crystal in stripes 3–6 μm wide and deep; these in-diffused regions have the higher index needed for them to act as waveguides.

We use the electro-optic effect in lithium niobate to change the guide's effective index [16] by applying an electric field transversely across the guide; a basic modulator is sketched in Figure 12.6. The transverse field comes from electrodes deposited on the substrate, and the propagation constant in the guide interacting with the field is proportional to the voltage.

Such technology, by analogy with that used for electronic integrated circuits, is called *integrated optics*. Integrated optics involves the processing of light waves by modulators, switches, and amplifiers in devices made (perhaps by different but compatible processes) on the same substrate. It could be used to process telephone signals carried optically in fiber-optic cables without having to translate them from light to electronics and back again.

Integrated optics is intended to miniaturize optics, make it rugged and insensitive to the environment, and make it inexpensive, which is why new optical gyros use integrated optics light processing elements.

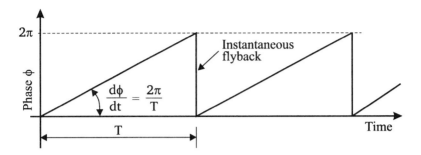

Figure 12.7. Serrodyne frequency shifting.

Serrodyne Frequency Shifting

Many IFOG builders use integrated optics phase modulators to close the loop, using the serrodyne frequency shifting approach originated by Arditty et al. at Thomson CSF. A serrodyne shifter is made from a phase modulator excited by a sawtooth voltage [17]. To make a frequency shifter with no sidebands, the phase is varied by 2π; the frequency generated depends on the rise time (T) of the sawtooth voltage, as shown in Figure 12.7. To see how this shifts frequency, let a represent the instantaneous amplitude of a light wave passing through the modulator:

$$a = a_0 \sin (2\pi ft + \phi)$$

where
  a_0 = wave amplitude
  f = optical frequency
  ϕ = phase

As we have chosen the slope of the phase ramp to be $2\pi/T$, the phase at any instant t is $2\pi t/T$. Therefore:

$$a = a_0 \sin [2\pi(f + 1/T)t]$$

which represents a wave with its frequency shifted by $1/T$. Note that instantaneously subtracting 2π doesn't affect the frequency shift. As $\Delta f = K\Omega$, then $\Omega = 1/(KT)$.

This method of frequency shifting is common because integrated optics modulators use little power and have relatively low insertion loss, but it makes demands on the electronics of the ramp driver. First, the ramp must vary in slope over the dynamic range of the gyro, always resetting at 2π phase. Second, any ripple in the rising voltage will generate frequency noise, which can end up at the

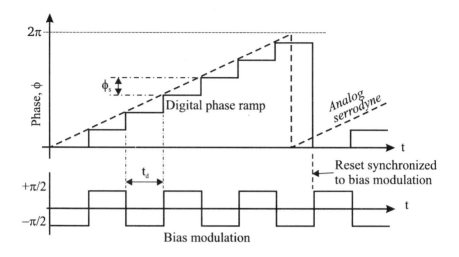

Figure 12.8. The digital phase ramp.

detector and raise the gyro threshold. Third, the flyback must be instantaneous to avoid gyro noise, and a finite flyback time will affect scale factor linearity.

Figure 12.8 illustrates the *digital phase ramp* approach [18], which was devised to improve the serrodyne scale factor linearity. Instead of a ramp, the modulator provides steps of phase with duration t_d, the transit time in the coil. These are reset synchronously with the bias modulation so that the flyback noise can be gated out. The output is a count of the number of phase steps between resets. References 18 and 19 describe this approach in more detail and show a schematic of the signal processing electronics.

Fiber-to-Chip Attachment—The JPL IFOG

Now that we have broken the fiber run to insert an integrated optics chip, we must solve the problem of connecting the fibers to the chip. The NASA/CalTech Jet Propulsion Laboratory solution [20] is typical; it uses a lithium niobate chip with two waveguide phase modulators on it, one for a serrodyne shifter and another for the quarter-wave bias (in lieu of the piezoelectric fiber stretcher). Others use the same elements in different designs.

Figure 12.9 shows how JPL attaches its fibers to the chip, using fibers polished at an angle, mating with the chip polished at a different angle, aligned in grooves etched in silicon mounts [21–23]. The end polishing is designed to minimize reflections at the joint (so that local interferometers are not formed) and maximize power transfer across the lithium niobate chip.

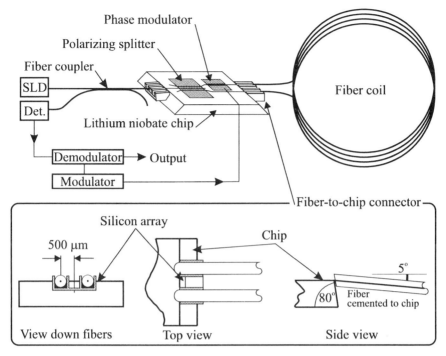

Figure 12.9. The Jet Propulsion Lab's IFOG.

Drift Due to Coil Temperature Gradients

We saw earlier how we can generate a bias by time delay modulation, purposely stretching the fiber with a piezoelectric driver. This same effect can also work against us, because a bias uncertainty (drift) can occur from a thermally induced non-reciprocity if there is a time-dependent temperature gradient along the fiber, due to temperature gradients across the coil. The nonreciprocity occurs, as it does in the piezoelectric phase bias device, when corresponding wavefronts of the counter-rotating beams cross the same region in the coil at different times. This effect is called the *Shupe effect*, after Shupe [24] who showed that the temperature gradients in a navigation-grade gyro must be less than 0.007°C for simple coil winding schemes.

The solution is to devise an *anti-Shupe* winding, to wind the coil so that parts of the fiber that are at equal distances from the coil center are beside each other [25–27]. One typical scheme is the quadrupole winding shown in Figure 12.10. "A" and "B" are two spools, each holding half the fiber for the coil. Spool A is attached to the coil former so that it rotates with it, and the first layers of the coil are wound from B. Then A and B change places; A is detached and B is fixed to

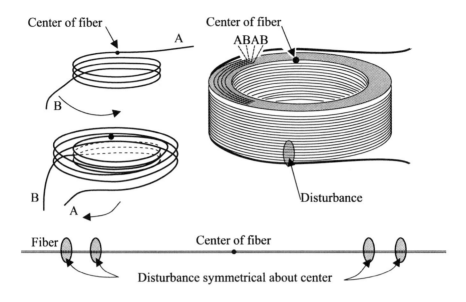

Figure 12.10. Anti-Shupe coil winding.

the coil former, and more layers are wound. This alternating process continues for the rest of the coil. In making the coil, the winder must be careful to minimize the distortion of the fibers as they cross one another as they pass between layers, because this causes microbends that increase fiber loss and scramble the state of polarization.

The Effect of Polarization on Gyro Drift

In Chapter 11 we described polarization maintaining (high-birefringence) fibers as having different propagation constants in orthogonal directions in the fiber plane. Modes traveling along the different directions have different effective indices, so that the fiber effective length differs for the orthogonal polarizations; waves accumulate an error phase from the passage through the alternate polarization. We must either fully randomize the phases (at some loss of sensitivity) or be sure that both the CW and CCW waves in an IFOG coil travel in the same polarization all the way around the fiber.

Consider an IFOG with a fiber coil of length L_{coil} in which there is a microbend or kink at a distance L_{cw} measured in the CW direction. Let the effective indices of the fiber for the two polarization modes be n_x and n_y. Light is launched in both directions with x polarization. The light traveling CW meets the microbend and some is converted from x to y polarization, so that the fiber effective length is

$$S_{cw} = L_{cw}n_x + (L_{coil} - L_{cw})n_y$$

Similarly, for the CCW wave

$$S_{ccw} = (L_{coil} - L_{cw})n_x + L_{cw} n_y$$

Then,

$$\Delta S = S_{cw} - S_{ccw} = (n_x - n_y)(L_{coil} - 2L_{cw})$$

Now the index difference is related to the fiber beat length, L [Equation (11.6)]:

$$n_x - n_y = \lambda_0/L$$
$$\Delta S = (L_{coil} - 2L_{cw})\lambda_0/L$$

If $L_{cw} = L_{coil}/2$, i.e., the defect is at the fiber midpoint, then $\Delta S = 0$. Generally ΔS is nonzero, so that there are two sets of coherent waves, each with the Sagnac phase shift, arriving at the detector:

1. the waves that traveled all the way in the x polarization, and
2. those that traveled part of the way in the y polarization.

They form two sets of overlapping fringes, and the detector senses a peak value offset from that due to set 1 alone, giving a gyro bias.

Although this is the essence of the polarization problem, this simplified view must be replaced by an analysis that allows for the polarization scrambling to be distributed along the fiber [28], and for the source coherence. The outcome, though, is unaltered: either use accurately randomized polarization (see, for example, [29]) or use strongly birefringent fiber (small h-parameter). In either case, wind it so that there are no defects, such as kinks, to disturb polarization or to cause cross-coupling.

If polarization cross-coupling occurs, the guided wave becomes elliptically polarized and, because of the Faraday effect, can develop a bias phase shift from magnetic fields along the fiber.

The Kerr Electro-Optic Effect

This Kerr effect can generate a nonreciprocal phase error so large that it swamps the Sagnac phase, which would make FOGs useless if there was not a way to minimize it. This effect describes the variation of a medium's refractive index under the influence of an electric field; it is a third-order nonlinearity. This means that the effective index of a fiber will depend on the optical power in its modes and will be given by $n_0 + \Delta n$, where n_0 is the propagation constant independent of power and Δn is the Kerr power-dependent term. For single-mode, polarization holding, fiber, carrying oppositely traveling waves of power P_{cw} and P_{ccw}, the differences in refractive indices are

$$\Delta n_{cw} = C(P_{cw} + 2P_{ccw})$$

$$\Delta n_{ccw} = C(P_{ccw} + 2P_{cw})$$

where C = constant, a function of the Kerr coefficient n_2. (These expressions are derived in many papers; Bergh et al. [30] gives an overview and bibliography.) The nonreciprocity arises from the factor of 2 in the right term, since the Kerr phase will depend on

$$\Delta n_{cw} - \Delta n_{ccw} = C(P_{ccw} - P_{cw})$$

Note that this is not a function of the absolute power, only the difference. If we let the difference in the Δns be represented by the propagation constant $\Delta\beta$ (Chapter 11), then:

$$\Delta\beta = 2\pi(\Delta n_{cw} - \Delta n_{ccw})/\lambda$$

and the Kerr phase is

$$\phi_K = \Delta\beta L$$

where L is the fiber length. From Equation (11.2), the Kerr-induced bias is

$$\Omega_K = \frac{\lambda c}{4\pi RL} \phi_K$$

Calculations show that for a typical navigation gyro, the Kerr-induced bias can be 0.2 deg/h/μW difference. However, by using a low-coherence source whose output is classed as "polarized thermal light," the optical field amplitude has a Gaussian probability distribution that statistically gives no net phase. The solution to the Rayleigh backscattering problem also solves the Kerr bias problem. Nature is rarely that cooperative.

The Fundamental Limit of IFOG Performance

There are two noise sources to consider for the IFOG (and any other sensor using a low-coherence source). The first is the shot noise, the second is the relative intensity noise (RIN).

IFOG Shot Noise

The fundamental limit (the noise "floor") of the IFOG is the photon shot noise, discussed in Chapter 11. Davis and Ezekiel [31] show that the minimum detectable rotation rate (threshold) is

$$\Delta\Omega = \frac{2}{K_o(n_{ph}\eta t)^{1/2}} \qquad (12.2)$$

where

n_{ph} = number of photons/s falling on the detector
η = detector quantum efficiency
t = averaging time for the measurement

and $K_o = 4dL/\lambda c$ is the open-loop scale factor [Equation (11.3)].

As the detector quantum efficiency is defined as the number of electrons emitted per photon on the detector, the current in the photodetector is $n_{ph}\eta e$, where e is the electron charge. The detector is more usually characterized by the responsivity, R, in A/W, and the power at the detector, P_d, is usually expressed in W. Accordingly,

$$n_{ph}\eta e = P_d R \qquad (12.3)$$

and (12.2) can then be written

$$\Delta\Omega = \frac{2}{K_o}\left(\frac{e}{P_d R t}\right)^{1/2} \qquad (12.4)$$

Let us calculate our model IFOG's shot noise limit, assuming the same gyro parameters as on page 191, i.e., L = 35,450 cm and d = 11.3 cm, λ = 1.3 µm, so that K_0 = 0.4 s. Further, assume that the power at the detector is 100 µW and that the detector has a responsivity of 0.5 A/W. The electron charge is 1.6×10^{-19} coulomb, so for 1-s samples, $\Delta\Omega = 2.75 \times 10^{-7}$ rad/s, or 0.06 deg/h. This would be a useful gyro for a heading and attitude reference system.

Relative Intensity Noise (RIN)

In the low-coherence (broad linewidth) source the spontaneous emission gives rise to many wavelengths beating against one another, causing intensity noise [32]. Because the coherence time is much less than the transit time around the coil the source noise cannot be subtracted from the Sagnac signal at the detector, as it can in resonant gyros with high-coherence sources. There has been a proposal to pass a sample of source noise through a fiber coil with the same delay time as the sensing coil so that this subtraction can take place [33], but this does not seem too practical.

Figure 12.11 illustrates how intensity noise passes to the detector. During bias periods a and b the noise in the detector output is uncorrelated and the noise level in Equation 12.4 increases [34] by the right-hand term:

$$\Delta\Omega = \frac{2}{K_o}\left(\frac{e}{P_d R t} + \frac{\lambda^2}{4c\Gamma t}\right)^{1/2}$$

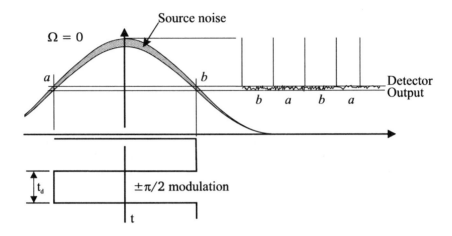

Figure 12.11. Relative intensity noise propagation.

The shot noise dominates if the detector power P_d is low, but for high power and large linewidth Γ the intensity noise dominates. Assuming for our model gyro that $\Gamma = 50$ nm the RIN contribution to gyro noise is 0.16 deg/h compared with the shot noise of 0.06 deg/h, so that the total rate uncertainty is 0.22 deg/h at t = 1 s.

Conclusions

The IFOG is being used in operational guidance and control systems, and is challenging the ring laser gyro (RLG). In fact, it has already displaced the RLG in some applications where gyro drift of 1 deg/h is acceptable. Boeing uses a Honeywell IFOG system as a backup attitude reference unit in its 777 airplane; Honeywell had shipped 1500 gyros by July 1996 [35]. Litton's LN-201 system (using IFOGs and micromachined silicon accelerometers) is being produced for the AMRAAM missile guidance system. Litton is also making a high-performance GPS-aided system under contract to DARPA. Note that the IEEE has recently prepared a standard for specifying and testing IFOGs [36].

We have described the problems they must overcome to make higher-performing gyros; in summary they are:

1. IFOGs must be biased for maximum sensitivity.
2. They must be operated closed-loop, which calls for a serrodyne shifter with complex electronics.
3. The attachment of the fibers to the integrated optics chip is difficult and must be carefully engineered.
4. The coil must be anti-Shupe wound.
5. Either the polarization must be rigorously randomized or the fiber must maintain polarization.

6. The light source must have a known stable wavelength, a broad line, e.g., 50 nm, and as high an output power as possible.

There are many benefits from solving these problems. The IFOG is lighter and smaller than the RLG, will probably have a longer life, and may be more rugged. Customers want optical gyros rather than mechanical gyros for strapdown systems, because optical gyros have very good high-rate performance, whereas mechanical gyros do not. The RLG blazed that trail by demonstrating high performance and very long life over thousands of systems.

There is a school of thought that believes that fiber-optic gyros can be less expensive if they use resonators rather than fiber coils, and others believe that a passive optical gyro can be made entirely on the integrated optics chip, again using a resonator. So in the next chapters we will consider the gyros made from optical resonators, first the active type, the RLG, then the passive versions.

References

1. Vali, V., R.W. Shorthill, reply to letter, Laser Focus, Nov. 1975.
2. Vali, V., R.W. Shorthill, "Ring interferometer 950 m long," Appl. Opt., 16, 2, pp. 290–291, Feb. 1977.
3. Ezekiel, S., H.J. Arditty (Eds.) *Fiber-Optic Rotation Sensors*, Springer-Verlag, New York, 1982.
4. Smith, R.B., *Selected Papers on Fiber-Optic Gyroscopes*, SPIE Milestone Series MS8, 1989. (570 references)
5. Davies, D.E.N., S. Kingsley, "Method of phase modulating signals in optical fibres: Application to optical-telemetry systems," Electron. Lett., 10, 21, 1974.
6. Davis, J.L., S. Ezekiel, "Closed loop, low noise fiber-optic rotation sensor," Opt. Lett., 6, 10, pp. 505–507, 1981.
7. Lefèvre, H.C., S. Vatoux, M. Papuchon, C. Puech "Integrated optics: A practical solution for the fiber-optic gyroscope," *Fiber-Optic Gyros: 10th Anniversary Conference*, Proc. SPIE 719, p. 101, 1987.
8. Böhm, K., P. Russer, R. Ulrich, E. Weidel, "Fibre-optic rotation sensor," DGON Symposium Gyro Technology, Stuttgart, 1980.
9. Cutler, C.C., S.A. Newton, H.J. Shaw, "Limitation of rotation sensing by scattering," Opt. Lett., 5, 11, pp. 488–490, Nov. 1980.
10. Hergenroeder, L., S.P. Smith, S. Ezekiel, "A chirped semiconductor laser as an alternative to the SLD in a fiber gyro," *Optical Fiber Sensors*, 1988 Technical Digest Series, Vol. 2, Optical Society of America, Washington, D.C., pp. 455–457, 1988.
11. Oho, S., H. Sonobe, T. Kumagai, H. Kajioka, S. Okabayashi, H. Nemoto, "Automotive navigation experiments with fiber gyroscope," *Optical Fiber Sensors*, 1988 Technical Digest Series, Vol. 2, Pt. 2, Optical Society of America, Washington, D.C., Paper PD3, 1988.

12. Wysocki, P.F., M.J.F. Digonnet, B.Y. Kim, H.J. Shaw, "Characteristics of Erbium-doped superfluorescent fiber sources for interferometric sensor applications," IEEE J. Lightwave Tech., 12, 3, pp. 550–567, March 1994.

13. Bergh, R.A., H.C. Lefèvre, H.J. Shaw, "All-single-mode fiber-optic gyroscope," Opt. Lett., 6, 4, pp. 198–200, April 1981.

14. Cahill, R.F., E. Udd, "Phase-nulling fiber-optic laser gyro," Opt. Lett., 4, 3, pp. 93–95, March 1979.

15. Yariv, A., "Guided-wave optics," Scientific American, 240, 1, pp. 64–72, Jan. 1979.

16. Papuchon, M, "Integrated optics," p. 116, and Leonberger, F.J., "Guided-wave electro-optic modulators," p. 130, in Ezekiel, S., H.J. Arditty (Eds.) *Fiber-Optic Rotation Sensors*, Springer-Verlag, New York, 1982.

17. Cumming, R.C., "The serrodyne frequency translator," Proc. IRE, p. 175, 1957.

18. Lefèvre, H.C., J.P. Bettini, S. Vatoux, M. Papuchon, "Progress in optical fiber gyroscopes using integrated optics," AGARD/NATO *Conference Report on Guided Optical Structures in the Military Environment*, AGARD CPP-383, 9A/1-13, 1985.

19. Lefèvre, H.C., *The Fiber Optic Gyroscope*, Artech House, Boston, 1993.

20. Goss, W.C., et al., "Closed loop fiber-optic rotation sensor," U.S. Patent 4 662 751, 5 May 1987.

21. Murphy, E.J., "Fiber attachment for guided wave devices," J. Lightwave Tech., 6, 6, June 1988. (71 references)

22. Minford, W.J., F.T. Stone, B.R. Youmans, R.K. Bartman, "Fiber-optic gyroscope using an eight-component LiNbO$_3$ integrated optic circuit," in Udd, E., R.P. De Paula (Eds.), *Fiber-Optic and Laser Sensors VII*, Proc. SPIE 1169, pp. 304–309, 1990.

23. Youmans, B.R., R.K. Bartman, P.M. Salomon, "Design and performance of a fiber-optic gyroscope using integrated optics," in Udd, E., R.P. De Paula (Eds.), *Fiber-Optic and Laser Sensors VII*, Proc. SPIE 1169, pp. 310–322, 1990.

24. Shupe, D.M., "Thermally induced non-reciprocity in the fiber-optic interferometer," Appl. Opt., 19, 5, pp. 654–655, 1 Mar. 80.

25. Frigo, N.J., "Compensation of linear sources of non-reciprocity in Sagnac interferometers," *Fiber-Optic and Laser Sensors I*, Proc. SPIE 412, pp. 268–271, 1983.

26. Bednarz, B., "Fiber-optic sensing coil," U.S. Patent 4 793 708, 27 Dec. 1988.

27. Arditty, H., et al., "Method of coiling an optical fiber gyroscope and an optical fiber coil thus obtained," U.S. Patent 4 743 115, 10 May 88.

28. Ulrich, R., M. Johnson, "Fiber ring interferometer: Polarization analysis," Opt. Lett., 4, 5, pp. 152–154, May 1979.

29. Blake, J.N., J.R. Feth, B. Szafraniec, "Configuration control of mode coupling errors," U.S. Patent 5 377 283, 27 Dec. 1994.

30. Bergh, R.A., H.C. Lefèvre, H.J. Shaw, "An overview of fiber-optic gyroscopes," J. Lightwave Tech., LT-2, 2, pp. 9–107, April 1984. (100 references)

31. Davis, J.L., S. Ezekiel, "Techniques for shot noise limited inertial rotation measurement using a multiturn fiber Sagnac interferometer," Proc. SPIE, Vol. 157, p. 131–136, 1978.
32. Taylor, H.F., "Intensity noise and spontaneous emission coupling in superluminescent light sources," IEEE J. Quantum Electronics, 26, 1, Jan. 1990.
33. Moeller, R.P., W.K. Burns, "1.06 µm all-fiber gyroscope with noise subtraction," Opt. Lett., 16, 23, pp. 1902–1904, 1 Dec. 1991.
34. Burns, W.K., R.P. Moeller, A. Dandridge, "Excess noise in fiber gyroscope sources," IEEE Photonics Technology Letters, 2, 8, Aug. 1990.
35. Klass, P.J., "Fiber-optic gyros now challenging laser gyros," Aviation Week and Space Technology, pp. 62–64, 1 July 1996
36. IEEE STD 952-1997. Specification Format Guide and Test Procedure for Single-Axis Interferometric Fiber-Optic Gyros.

13
The Ring Laser Gyro

In Chapter 11 we defined the ring laser gyroscope (RLG) as an active resonator optical gyro. Clifford Heer conceived the RLG in 1961 [1]; he saw that the properties of the laser, recently invented by Schawlow and Townes, could be exploited to measure rotation. Heer and Adolph Rosenthal [2] independently developed the theory, and, in 1963, Macek and Davis [3] demonstrated the first RLG, a square gyro, 1 m on a side. Scientists around the world contributed to the field during the 1970s, and Bogdanov's survey article [4] describes the results.

Modern aircraft navigators and heading and attitude reference systems (HARS), use RLGs with perimeters in the 15–30-cm range, and those for tactical guidance systems have about 10 cm perimeter. In this chapter we will first look at the principle of the gas laser, describe a problem called *lock-in,* which must be overcome, and then examine the components of RLGs by using different companies' designs as case studies. Before proceeding we should note that Aronowitz [5] has written a treatise on the RLG's basic operation that is invaluable to the interested student.

The Laser

The laser is an optical oscillator. A laser (the word is an acronym for Light Amplification by Stimulated Emission of Radiation) consists of an amplifying medium inside an optical resonator. The medium is made to amplify by some excitation; in gas lasers this may be an electrically energized plasma, whereas in semiconductors an electric current excites electrons into a higher energy band. Light can either be emitted spontaneously by an excited medium, or it can be stimulated to emit, a condition that leads to coherent monochromatic radiation.

Stimulated Emission

Atoms are excited by adding energy quanta to their orbiting electrons. The excited electrons can drop to a lower energy excited state at any time, emitting a photon

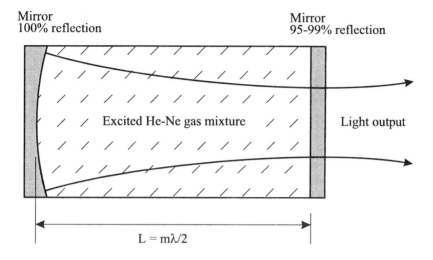

Mirror
100% reflection

Mirror
95-99% reflection

Excited He-Ne gas mixture

Light output

$L = m\lambda/2$

Figure 13.1. A Fabry-Perot gas laser.

characteristic of the transition between the upper and lower excited states [6]. This happens in the neon lamp; the characteristic orange glow comes from a neon atomic transition. Such light is incoherent; the phase of the emitted light varies randomly because the atoms emit their photons spontaneously, at any time.

In 1917 Einstein introduced the theory of stimulated emission. An atom in an excited state may be induced to emit a photon at a particular time by the presence of radiation, in which case its emission is coherent, because the emitted light picks up the phase of the stimulating radiation. To make a laser, it is necessary to generate stimulating radiation at the expense of the spontaneous emission.

In order for a gas to behave as an amplifying medium it must be excited in such a way that there are more atoms in the excited state than in the unexcited state, a condition called *population inversion*. This situation does not occur easily; it can be achieved in neon by mixing it with helium, because then the helium becomes excited and passes energy onto the neon and the laser transition occurs in the neon atom.

We place this excitable gas in a resonator (e.g., a Fabry-Perot, Chapter 11), as shown in Figure 13.1, and excite the gas. An atom will spontaneously emit a photon, which will stimulate emission from a nearby atom, which stimulates emission from another nearby atom, and so on in an avalanche. But some of the emitted photons hit the walls, some are absorbed by atoms temporarily in the unexcited state, and some of the atoms are traveling at the wrong speed. Their emitted frequency will be Doppler shifted so that their energy will not match that needed by the target atom.

If the gain provided by the amplifying medium exceeds these losses, waves will be generated in both axial directions, so that a standing wave builds up in the

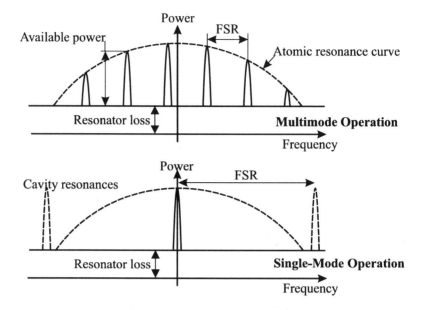

Figure 13.2. Gas laser longitudinal modes.

resonator. Its wavelength will be such that an integral number of half-wavelengths fit into the resonator, and its amplitude will be limited by the resonator losses. The gain from a gas laser is not very large, so one cannot take very much power from the resonator; typical helium-neon lasers emit a few milliwatts of light at a wavelength of the order of 1 µm. To make a high-power gas laser one must make it large, up to a meter long.

A resonator resonates when its length is an integral number of half-waves, but oscillation can occur only at those frequencies where the excited atom has excess gain, as shown in Figure 13.2. The supported frequencies are called *longitudinal modes*. The atomic resonance curve (*gain curve*) shows the power available as a function of frequency, its frequency width being called the *gain width*. The number of longitudinal modes possible depends on the gain width and the length of the resonator, and the laser is usually tuned so that one longitudinal mode falls at the peak of the gain curve, thus maximizing output. To do this one can move one of the mirrors axially using a piezoelectric mount, for it is only necessary to move the mirror 0.5 µm or so because one half-wavelength will change the frequency by one free spectral range (FSR).

The allowed transverse modes depend on the curvature and spacing of the mirrors and the losses for the mode; gyro lasers operate in the lowest-order transverse mode. HeNe lasers can emit light at 0.63 µm (in the red), 1.15 µm (near infrared), and at 3.39 µm.

Gas lasers emit highly coherent light with a small beam divergence, but they need high voltages (600–1000 V) to excite the gas. They are large, of the order of 2 to 20 in. Their mirrors must be accurately and stably positioned to a few arc-

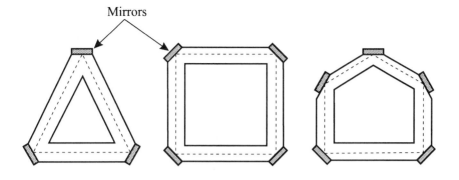

Figure 13.3. Ring resonators with three, four, and five mirrors.

seconds. The gas fill must be free of contamination, which would poison the cathode or deposit (by sputtering) on the mirrors; they are made with vacuum tube technology. As a result, they cost hundreds of dollars.

Vacuum electron tubes limited the usefulness of electronics, a field that burst out in myriad ways when semiconductor devices provided integrated circuits. In the same way, the gas laser was limiting coherent optics development until the development of the semiconductor laser liberated the field of opto-electronics. Now, it, too, is flourishing—the compact disk player is an obvious example.

The Semiconductor Laser

Solid-state lasers can be made in semiconductors [7]. Lasers for CD players are made from gallium aluminum arsenide diodes, which are cubes about 250 µm on a side—tiny crystals the size of a grain of sugar. Instead of being placed in a mirror resonator, the GaAlAs crystal facets themselves form mirrors that cause the light to oscillate in the amplifying medium. This medium, the *active region*, is a thin layer (about 0.3 µm thick) of different composition where the current passing through the diode causes the population inversion and provides gain. Gallium aluminum arsenide diodes emit light at around 0.83 µm; other alloys provide different wavelengths. The intensity and frequency of the light are a function of the current through the diode, so they are readily controlled. Diode lasers emit an enormous power for their size; they can provide 100 mW in a single mode at 0.82 µm. However, the output light diverges (from diffraction at the output of the active region) into a wide-angled cone with an elliptical base, requiring astigmatic optics for coupling their output into waveguides and fibers.

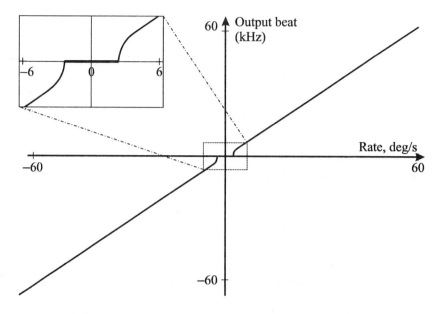

Figure 13.4. RLG lock-in.

The Ring Laser

One can make a Fabry-Perot resonator with three or more mirrors, making the light circulate through the gain medium, as in Figure 13.3. Both CW and CCW waves will be generated, which will resonate when the path perimeter is an integral number of wavelengths, and the two waves form a standing wave pattern. Such a laser is called a *ring laser* and is an active resonator in the terminology of Chapter 11; its FSR is given by Equation (11.10). As described earlier, the number of longitudinal modes supported depends on the FSR and the gain width of the lasing transition's atomic resonance curve. The width of the HeNe 0.63-µm transition gain curve is about 2 GHz, so if a single longitudinal mode is to be excited, the ring perimeter must be less than 1 m under useful excitation conditions.

There is another way of looking at the Sagnac effect, different from our approach in Chapter 11. Consider that the resonator standing wave pattern is the RLG's inertial reference; it remains stationary while the world moves about it. There may be as many as 10^6 wave periods in a typical RLG, and the gyro's optical detectors sense the intensity of the light at a point in the standing wave. They can therefore interpret output intensity as case rotation relative to the standing wave, with a resolution of the order of a millionth of the perimeter, 1 arc-sec or so.

Lock-In

Reverting to the view of the Sagnac effect described in Chapter 11, the RLG's CW and CCW waves should have different frequencies when we rotate the gyro about an axis normal to the resonator plane. However, the expected output frequency does not appear at low rotation rates. The output stays at zero until a critical rate is reached, when it jumps close to the expected value. The CW and CCW resonances lock together until the frequency difference is large enough to separate them. This characteristic, shown in Figure 13.4, is called *lock-in*. It is caused by the backscattered light in the resonator coupling the two beams in the lasing medium.

It is necessary to create a frequency difference, a real or artificial rotation rate (rate bias), to overcome lock-in. Four biasing approaches have been tried. The first, most obvious, approach is continuous rotation; rotation at say 50 deg/s, reversed every two revolutions, gives good results [8], although at considerable mechanical cost. The second, and most used, approach is *mechanical dithering*; the third uses a *magnetic mirror*; and the fourth uses a DC bias in a *multioscillator* gyro. We will consider these last three approaches by studying RLG designs from the different companies that have specialized in those fields.

Mechanical Dither

In the mechanically dithered gyro the resonator is oscillated in its plane (about its input axis, IA) by a piezoelectric motor [9], at an angular rate high enough to overcome lock-in for most of the dither cycle. The motor is driven at the resonant frequency of the block on the motor, and the block is dynamically balanced. This avoids two problems; it prevents dither vibration from shaking the case excessively, and vehicle vibrations (or dither vibration from other gyros) will not couple into the gyro motion and disturb the IA direction.

At the extremes of the dither motion the angular velocity drops to zero, so that the CW and CCW beams lock momentarily. If the dither is sinusoidal, the gyro experiences constant end-point lock-in conditions, which are generally not symmetrical at the opposite extremes of the dither. As a result there is a net bias with a steady-state value of the order of 10 deg/h; the value depends on the lock-in threshold and cannot be compensated.

To overcome this problem, the dither drive frequency has a random frequency component superimposed on it so that the block motion is slightly randomized; now the motion has a random rate noise instead of a mean bias. This [as with shot noise, Equation (11.5)] causes a random walk in angle, given by [10],

$$R_{\theta D} = f_L \left(\frac{t}{2\pi K f_m} \right)^{1/2}$$

(13.1)

where
    f_L = lock-in threshold

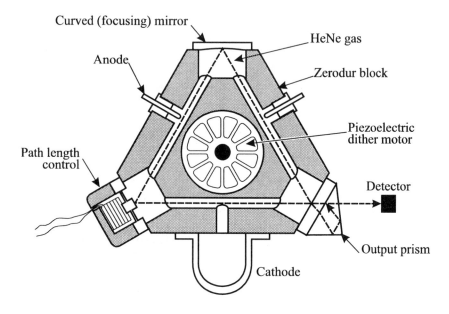

Figure 13.5. A dithered RLG.

f_m = peak dither amplitude
K = scale factor

This dither noise is distinct from the spontaneous emission (quantum) noise. One key advantage of mechanical dithering is that it cannot cause a long-term angle error (assuming that no output counts are lost), for even though the dither causes noise, there can be no net rotation of the mechanical assembly.

Podgorski (Honeywell) was awarded a patent for the integrated RLG [11] in 1969; it is shown schematically in Figure 13.5. An ultra-low-expansion glass/ceramic ("Zerodur" or "Cervit") block carries three mirrors on precisely polished faces, aligned to arc-second accuracy so that a light beam will pass around the path without wandering off its axis, thereby forming a stable resonator. A cathode and two anodes are sealed to the block, which is filled with a helium-neon gas mixture. One mirror is made so that it can be translated by a piezoelectric driver to change the optical perimeter; it is used to tune the resonator, driven by a servo that maintains the laser intensity at its peak.

The laser is excited by generating a plasma in the gas, which is ionized by applying a few hundred volts between the cathode and anodes. When the discharge occurs, ions flow along the laser path (Langmuir flow) and give rise to Fresnel drag, a source of bias. Splitting the discharge (by providing the two anodes) generates two opposite flows whose biases cancel, a technique used in all RLGs. (If Fresnel drag generates a bias, why not use it to bias a gyro against lock-in? Some tried, but found it insufficiently stable.)

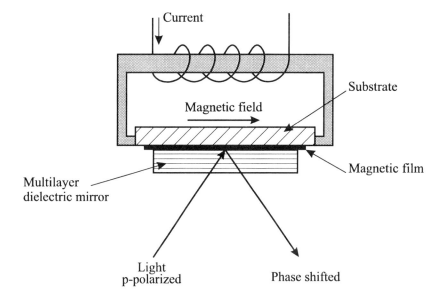

Figure 13.6. The Kerr-effect magnetic mirror.

Samples of the CW and CCW waves leave the resonator through a partially transmitting mirror and are combined in a prism so that they meet on a detector that generates the output beat frequency.

With the integral laser design, the mirrors are exposed to the plasma ion bombardment, which can degrade their performance. Developing mirrors "hard coated" to survive in the plasma was a key achievement. Killpatrick described the Honeywell gyro in more detail [12]; by the end of 1997, Honeywell had delivered more than 100,000 of these gyros. Honeywell strapdown navigators have consistently performed to better than 1 nm/h and have demonstrated higher reliability than mechanical gyro systems. A Japanese group has published an interesting paper [13] describing their development of a Honeywell-style RLG.

Kearfott and other companies in the United States and overseas developed similar mechanically dithered triangular RLGs, whereas Litton used a resonator with four mirrors; Northrop [14] used five, partly because backscatter (and hence lock-in) is lower with greater angles of incidence on the mirrors.

The Magnetic Mirror

The rise of Honeywell and the fall of Sperry in the RLG business can be connected to their design dogmas. Macek and Davis of Sperry demonstrated the first RLG, and that company embarked on a development that seemed in the late 1960s to be the way to go. Sperry decided to overcome lock-in with an optical bias device, the

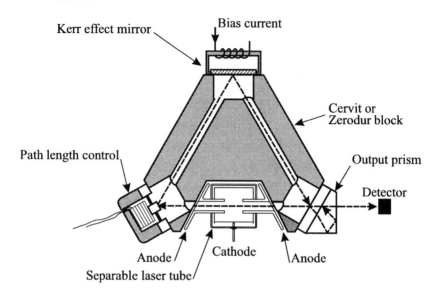

Figure 13.7. A Sperry-style RLG.

magnetic mirror, which has no moving parts. The magnetic mirror has the advantage over mechanical dither that, in an IMU, the gyros do not vibrate the accelerometers and one another, a potential source of system error.

Magnetic mirrors use the magneto-optic Kerr effect [15] to produce a non-reciprocity in the resonator, illustrated in Figure 13.6. This effect relates to the phase change produced in a wave when it is reflected at the surface of a magnetized ferromagnetic substance [16, 17], a phase change that depends on the alignment of the incident wave and the magnetizing field. If one mirror of the gyro has a ferromagnetic film deposited on it that is magnetized in the plane of the resonator, then the CW wave will experience an opposite phase shift from the CCW wave. This phase difference will translate to a frequency shift in each wave, generating a bias that takes the gyro out of the lock-in range. Commutating the bias by reversing the direction of magnetization (reversing the polarity of the current in the electromagnet generating the field) avoids long-term drifts.

Now Sperry considered the integration of the laser and cavity to be difficult to make reliable. It was necessary to make seals at the mirrors and the electrodes and to make the cavity so clean that there would be no outgassing to contaminate the lasing gas. Sperry had experience in the field of electronic vacuum tubes, so they separated the laser from the cavity by treating the laser as a vacuum tube like a magnetron. They designed a separable, field-replaceable laser tube. As a bonus, this approach isolates the resonator mirrors from the plasma, so they can be "soft coated," a process that in the early 1970s gave a higher yield of high reflectivity mirrors. Also, three magnetic mirror gyros could be integrated into one block with three laser tubes, making their system small and rugged.

Sperry knew how to make glass envelopes, cathodes, and anodes; how to seal, bake out, and backfill, and saw this approach as superior to Honeywell's dithered monolithic design. In those days Honeywell was mired in the technology problems that it was to solve so successfully, but no one knew that then. Sperry made two sizes of gyro, the SLIC 7 and 15 [18,19], for tactical and aircraft-grade navigation respectively; their gyros looked something like Figure 13.7.

Magnetic mirrors appear attractive, but they have a few snags. The presence of the ferromagnetic film in the mirror increases its loss, which increases the lock-in threshold, the very thing you do not want, because then you have to increase the Kerr bias. The magnetizing field had to be quite large in Sperry's design, requiring a field coil with such a high inductance that it could not be switched quickly. This meant that the bias could only be alternated at a few Hertz, not as often as one would prefer, as it limits bandwidth.

The gyro performance depends on the resonator finesse (Chapter 11); the higher the finesse the better. Because the inclusion of the Kerr effect film increases optical loss, the dithered gyro has lower-loss mirrors than the magnetic mirror version. Also, Sperry's separable laser must have windows between the tube and the block containing the mirrors, and these windows cause loss and scatter, reducing finesse (and increasing the lock-in band). Once the technology was developed, the dithered gyro turned out to be better value for money.

During the 1980s, Northrop developed a tactical RLG [20, 21] designed for low-cost production. It used a new type of magnetic mirror based on improvements in magnetic materials from the recording industry and monolithic construction with plasma-resistant mirrors. It did not go into production.

The Multioscillator

In the multioscillator RLG, two gyros are built into the same block, arranged so that the same DC bias is applied to each of them in opposite senses, so that by adding the outputs of the gyros the common bias is eliminated. This kind of gyro is variously known as a *four-frequency gyro* (two frequencies and two gyros) or as a DILAG, a *differential laser gyro* [22]; both Raytheon and United Technologies hold patents [23,24]. Neither company currently sells one; Raytheon's technology is marketed by Litton as the ZLG (zero-lock gyro).

The multioscillator concept is shown in Figure 13.8, which shows a four-mirror gyro with a lasing medium and two other optical elements, a Faraday rotator and a reciprocal polarization rotator. The lasing medium is the composite gain curve from the two isotopes of neon. The reciprocal polarization rotator is a piece of crystalline quartz aligned with its optical axis along the direction of light propagation. It is birefringent for circularly polarized light, i.e., it has a lower refractive index to one direction of propagation and higher index to the other. This makes the gyro resonate only in circularly polarized modes, and the birefringence provides a bias between the right- and left-circularly polarized (RCP and LCP) resonances, chosen to be the free spectral range, $c/2p$, as shown in Figure 13.9.

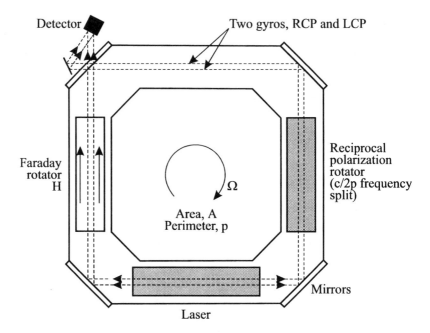

Figure 13.8. The multioscillator (four-frequency) RLG.

The Faraday rotator consists of a piece of glass in an axial magnetic field. When circularly polarized light passes along the axis in the glass, the wave is retarded if it travels opposing the magnetic field, and vice versa. Accordingly, the Faraday rotator provides a DC bias for both the LCP and RCP resonances in the gyro. Thus

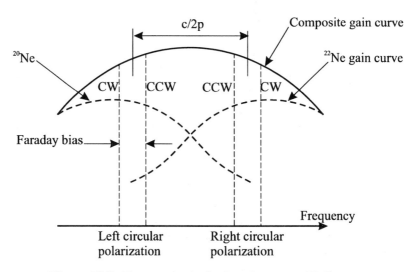

Figure 13.9. Frequencies in the four-frequency RLG.

there are four waves circulating, distributed as shown in Figure 13.9. With the refractive index in the cavity assumed to be unity, taking the output of the LCP gyro,

$$\Delta f_L = K_1 \Omega + B$$

where
   K_1 = two-frequency RLG scale factor [Equation (11.13)]
   B = bias

and the output of the RCP gyro,

$$\Delta f_R = K_1 \Omega - B$$

and summing them, we get twice the scale factor of a two-frequency RLG:

$$\Delta f = 2K_1 \Omega$$

Although the multioscillator seemed like a low-noise solution to the lock-in problem, early models did not perform well enough to compete in their cost bracket. The losses due to components in the beam path caused backscattering and lower finesse, and the losses changed with temperature as the active elements shifted position by small amounts. Later, Raytheon invented the out-of-plane resonator in which they replaced the quartz crystal by a cunning geometrical construction [25]. This provided the necessary π phase between the RCP and LCP waves, considerably improving performance. Litton purchased the technology and sells it as the *zero-lock gyro*, ZLG.

Shared-Mirror RLG Assemblies

Simms at British Aerospace seems to be the inventor of a neat "three-in-one" package in which three orthogonal RLGs share six mirrors and a single dither motor. We call this the "shared-mirror" configuration.

   Honeywell and Sperry built three-axis gyro assemblies as early as 1967, making three separate gyros in one block. If the three gyros are on the adjacent surfaces of a cube, then mechanical dither bias can be provided by oscillating (or rotating) the cube around the diagonal through the corner where the faces meet. In the BAe design, six mirrors are shared between three square orthogonal gyros as shown in Figure 13.10 [26, 27]. Such an assembly must still be dithered, but it might be a cost-effective solution for lower-accuracy navigators, providing that the interactions between the gyro axes caused by mirror scattering from one axis to another can be minimized. The three gyros can use a common cathode and can contain a reservoir of gas for longer life. Kearfott has developed a version of this concept that carries the "TRILAG" trademark; the T24 model [28] has

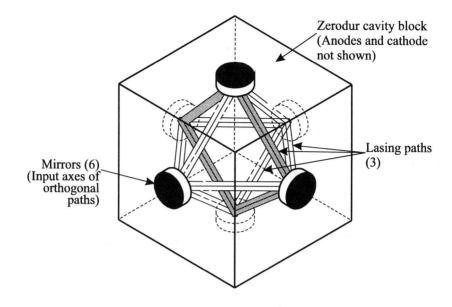

Zerodur cavity block
(Anodes and cathode
not shown)

Lasing paths
(3)

Mirrors (6)
(Input axes of
orthogonal
paths)

Figure 13.10. Shared-mirror three-axis RLG.

demonstrated 1.5 m/h in a system, and the smaller T16 gyro performs in the 0.1-deg/h range. SFENA (France) and Litton also have patents for shared-mirror RLGs.

The Quantum Fundamental Limit

The RLG's background noise from the ring laser spontaneous emission gives an uncertainty in rotation rate [29]:

$$\Delta\Omega = \frac{\Gamma}{K(n_c t)^{\frac{1}{2}}}$$ (13.2)

where
Γ = resonator linewidth
K = gyro scale factor
n_c = number of photons/s in the resonator
t = averaging time

The resonator linewidth here is the "cold cavity" linewidth, that without the laser operating, the same value as that used in finesse calculations. The laser power in the resonator, P_c, is

$$P_c = n_c h\nu$$

where $h\nu$ = quantum energy. With F = finesse, free spectral range f_{sr} = c/p, and $\Gamma = f_{sr}/F$,

$$\Delta\Omega = \frac{c(h\nu)^{1/2}}{pKF(P_c t)^{1/2}} \tag{13.3}$$

This white spectrum rate noise causes a random walk in angle, just like the dither noise [Equation (13.1)]. Hammons and Ashby [10] provides an equation for this quantum random walk:

$$R_{\theta Q} = \frac{K'\lambda c}{8\pi A}(\chi h\nu/P_c)^{1/2} \quad rad/\sqrt{s} \tag{13.4}$$

where
  K' = constant
  A = resonator area
  χ = loss per round trip

K' represents the noise contribution from the nonempty lasing level; it has the approximate value of 1.67. Also, 1 rad/$\sqrt{s}$ = 3.438×10^3 deg/$\sqrt{h}$, and the loss per roundtrip and the finesse are related by $F = 2\pi/\chi$.

We can calculate $\Delta\Omega$, $R_{\theta Q}$, and $R_{\theta D}$ for a typical navigation-grade gyro, which we will assume to have an equilateral triangular light path of perimeter 30 cm. Its area A = 43 cm². The laser is using the 0.63-µm transition whose quantum energy $h\nu = 3.14 \times 10^{-19}$ J. A good RLG can have a finesse of 5000, from which we can calculate the resonator linewidth and scale factor:

$\Gamma = c/pF = 2 \times 10^5$ Hz
K = 4A/λp = 0.9 × 10^5 pulse/rad

The pulse weight (1/K) is 2 arc-sec/pulse. Therefore, with 1 s sampling (t = 1 s), Equation (13.3) gives the minimum detectable rate:

$\Delta\Omega = 4 \times 10^{-8}$ rad/s = 0.008 deg/h

For a finesse of 5000, the loss per roundtrip is 0.0013, and the quantum random walk in angle [Equation (13.4)] is

$R_{\theta Q}$ = 0.006 deg/$\sqrt{h}$

To estimate the dither random walk, assume that the lock-in threshold is 30 deg/h and that the peak dither amplitude is 100 deg/s. Then Equation (13.1) gives

$R_{\theta D}$ = 0.0038 deg/$\sqrt{h}$

In this example the mechanical dither noise does not limit the gyro performance, because it is less than the quantum noise.

Quantization Noise

The gyro output is in pulses and is sampled periodically. At most rotation rates there will not be an exact integer number of pulses in the sample period, so that some remainder will carry into the next period. The electronics must not lose this remainder; the output must be read "on the fly" while the pulse counting continues. However, the remainder is the source of quantization noise. Assume that we rotate the gyro at a steady rate Ω. The number of pulses generated in a period t is

$$N = \Omega t/S$$

where S = pulse weight, arc-sec/pulse. The whole number counted, P, is the integer value of N, and the remainder $R = N - P$. Therefore we will get P counts for most samples, but we will get an extra pulse every 1/R periods. The standard deviation of this process can be shown to be

$$\sigma = (S/t)(R(1-R))^{\frac{1}{2}}$$

This is zero when $R = 0$ or 1 (when there is no remainder) and has its highest value when $R = 0.5$:

$$\sigma_{max} = S/2t$$

This can be a large error; for our model gyro with S = 2 arc-sec/pulse, the quantization error is 1 deg/h with 1-s samples.

There are two ways to reduce this error. First, design the readout hardware so that two detectors look at the fringe pattern and detect zero-crossings rather than just maxima. This reduces the pulse weight to S/4. Then special software techniques can be used to filter the quantization from the random walk [30].

The IEEE has released a specification guide, model equation, and test procedure for the RLG [31].

Conclusion

The SDF mechanical gyro still dominates the highest-precision strategic missile guidance field, although it is experiencing a serious challenge from the RLG, particularly from the Litton multioscillator ZLG. Platforms in which the RLGs are continuously rotated (thus avoiding the drift noise from the mechanical dither) have been shown to be close to the performance necessary for ICBM guidance.

Quantization error begins to be important as RLG size decreases, whereas passive optical gyros do not inherently suffer from quantization noise. IFOGs are challenging RLGs in the 0.01- to 1-deg/h range, where scale factor can be allowed errors up to 100 ppm, and IFOGs are well-suited to GPS-aided systems. But where better performance is needed, the RLG is preferred.

Currently, the dithered RLG is firmly embedded in the 1-m/h navigator business but it will come under cost and reliability pressure from the IFOG, since the RLG high-voltage supply is a cause for concern. Only time will tell which design approach will dominate the tactical (1-deg/h) market, where low cost is of dominant importance. The RLG may be shock hardened so that it can be used in cannon-launched projectile guidance, although there, again, it competes with the interferometric fiber-optic gyro and micromachined vibrating gyros.

References

1. Heer, C.V., "Interference of electromagnetic and matter waves in a nonpermanent gravitational field," Bull. Am. Phys. Soc. 6, 58, Jan. 1961.
2. Rosenthal, A.H., "Regenerative circulatory multiple-beam interferometry for the study of light propagation effects," J. Opt. Soc. Am., 52, 10, pp. 1143–1148, Oct. 1962.
3. Macek, W.M., D.T.M. Davis, "Rotation rate sensing with traveling-wave ring lasers," Appl. Phys. Lett., 2, 3, pp. 67–68, 1 Feb. 1963.
4. Bogdanov, A.D., "Laser gyroscopes," NASA TT F-15, 641, June 1974.
5. Aronowitz, F., "The laser gyro," *Laser Applications*, Monte Ross (Ed.), Volume 1, Academic Press, New York, 1971.
6. Fowles, G.R, *Introduction to Modern Optics*, Holt, Rinehart, and Winston, Inc., New York, 1968.
7. Yariv, A., *Introduction to Optical Electronics*, Holt, Rinehart, and Winston, Inc., New York, Chapter 7, 1971.
8. Matthews, A., H. Welter, "Cost-effective, high-accuracy inertial navigation," Navigation, J. Inst. Nav., 36, 2, pp. 157–172, Summer 1989.
9. Shackleton, B.R., "Mechanical design considerations for a ring laser gyro dither mechanism," *Mechanical Technology of Inertial Devices*, Proc. Inst. Mech. Eng. (London), 1987-2, Paper C60/87.
10. Hammons, S.W., V.J. Ashby, "Mechanically dithered RLG at the quantum limit," IEEE NAECON Record, Vol. 1, pp. 388–392, 1982.
11. Podgorski, T.J., "Control apparatus," U.S. Patent 3 390 606, 2 July 1968.
12. Killpatrick, J., "The laser gyro," IEEE Spectrum, pp. 44–55, Oct. 1967.
13. Tahara, Y., Y. Takizawa, H. Nakayasu, M. Otsuki, Y. Kojima, Y. Uehara, "Development of a space-use ring laser gyroscope in Japan," DGON Symposium Gyro Technology, Stuttgart, 1985.
14. Lim, W.L, J.P. Hauck, J.W. Raquet, "Pentagonal ring laser gyro design," U.S. Patent 4 705 398, 10 Nov. 1987.
15. Kerr, J., "On rotation of the plane of polarization by reflection from the pole of a magnet," Phil. Mag., 3, ser. 5, pp. 321–342, May 1877.

16. Krebs, J.J., W.G. Maisch, G.A. Prinz, D.W. Forester, "Applications of magneto-optics in ring laser gyroscopes," IEEE Trans. on Magnetics, MAG-16, 5, pp. 1179–1184, Sept. 1980.

17. Gauert, R., "Investigations of magneto-optic bias elements for laser gyros," DGON Symposium Gyro Technology, Stuttgart, 1981.

18. Morrison, R.F., C.B. Strang, "A missile laser gyro rate sensor," AIAA Guidance and Control Conference, 16–18 Aug. 1976.

19. Morrison, R.F., E. Levinson, R.W. McAdory, "The SLIC-15 laser gyro IMU for midcourse missile guidance," Institute of Navigation, April 1976.

20. Hughes, D., "Northrop develops miniature laser gyros for tactical missiles," Aviation Week and Space Technology, pp. 77-78, 8 Feb. 1988.

21. Perlmutter, M.S., J.M. Bresman, H.A. Perkins, "A low cost tactical ring laser gyroscope with no moving parts," DGON Symposium Gyro Technology, Stuttgart, 1990.

22. Bresman, J., H. Cook, D. Lysobey, "Differential laser gyro development," Navigation, J. Inst. Nav., 24, 2, pp. 153–159, Summer 1977.

23. Andringa, K., "Laser gyroscope," U.S. Patents 3 741 657, 26 June 1973, and 3 854 819, 17 Dec. 1974.

24. Yntema, G.B., D.C. Grant, R.T. Warner, "Differential laser gyro system," U.S. Patent 3 862 803, 28 Jan. 1975.

25. Dorschner, T.A., I.W. Smith, "Clear-path four-frequency resonators for ring laser gyros," J. Opt. Soc., 68, p.1381, 1978.

26. Simms, G.J., "Ring laser gyroscopes," U.S. Patent 4 407 583, 4 Oct. 1983.

27. Simms, G.J., "A tri-axial laser gyro," *Mechanical Technology of Inertial Devices*, Proc. Inst. Mech. Eng. (London), 1987-2, Paper C56/87.

28. Weber, D.J., "A three-axis monolithic ring laser gyro," Navigation, J. Inst. Nav., 35, 1, pp. 15–22, Spring 1988.

29. Ezekiel, S., H.J. Arditty (Eds.) *Fiber-Optic Rotation Sensors*, Springer-Verlag, New York, p. 7, 1982.

30. Mark, J., A. Brown, "Laser gyroscope random walk determination using a fast filtering technique," DGON Symposium Gyro Technology, Stuttgart, 1984.

31. IEEE STD 647-1981. Specification Format Guide and Test Procedure for Single-Axis Laser Gyros.

14
Passive Resonant Gyros

In Chapter 11 we divided resonator gyros into active and passive types, and in Chapter 13 we described active ring laser gyros, RLGs. Now we describe the passive resonator gyros, which can be made from either "free space" optics—lenses, mirrors, and other discrete components—or from guided wave optics—fibers or integrated optics, waveguides, and thin film devices. The free-space type might use an empty RLG-like block with mirrors on it, while the guided wave type can use an optical fiber resonator or an integrated optics resonator in which waveguides are formed into an optical chip. We shall look at each in turn, and then we'll compare them.

The Discrete Component Passive Ring Resonator

Ezekiel and Balsamo of MIT first described a practical passive ring resonator gyro (PARR) [1–3], sketched in Figure 14.1. Four mirrors define a resonator, whose CW and CCW resonances are probed by waves originating in a helium-neon laser (long coherence source, Chapter 11). Each wave has its frequency altered by a bulk acousto-optic Bragg frequency shifter, which adds (or subtracts, depending on beam entry angle) whatever RF excites a bulk acoustic wave in the Bragg cell. This wave diffracts the light beam, Doppler shifting its frequency. Detectors measure the intensity of the light transmitted through the resonator, which is maximum at resonance, and servos tune the cavity perimeter and one frequency shifter to hold on to both resonances. The second shifter is driven by a fixed RF signal, and the gyro output is the difference between the shifter frequencies. Because the laser has a coherence length greater than the optical path length, intensity noise, which we saw in Chapter 12, degraded IFOG resolution, can be subtracted from the difference signal and poses no problem.

This PARR uses free-space light propagation (rather than guided wave), making it difficult to control the transverse modes in the mirror-defined passive ring, and careful design and assembly are needed to operate in the fundamental TEM_{00} mode in the chosen polarization, necessary for stable bias. Because the resonator must be excited from outside, the resonance-probing laser beams must enter and leave

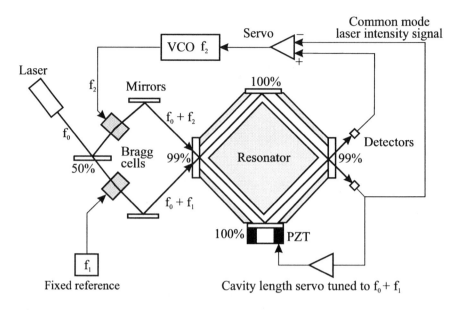

Figure 14.1. Ezekiel's passive ring resonator gyro.

through the mirror substrates, paths in glass that may be stressed and thus birefringent, causing polarization rotation that varies with time and temperature.

Another problem is that the light from the laser, upon division at the first beam splitter, passes by two separate paths through the Bragg cells, much as in a Mach-Zehnder interferometer. Any temperature gradients or air-pressure gradients (e.g., sound waves) passing across the gyro affect the cells at different times, causing fluctuations in optical path length (path and/or refractive index), which result in frequency noise. All laser gyros rely on common mode rejection to cancel environmental noise occurring in the CW and CCW paths, so these paths must be designed to be coincident. The RLG paths are about as coincident as they can be and the interferometric fiber-optic gyro (IFOG) manages particularly well by combining them in a fiber, but the discrete component PARR must separate the paths to pass through independently variable frequency shifters, and so has poorer common mode rejection.

It is interesting to note that Ezekiel's MIT group has measured lock-in in a PARR [4]. The amount was less than in an RLG; so far, no one else has reported this effect.

If one is prepared to pay the fabrication costs of a PARR, one might as well seal it, fill it with gas, add electrodes and a dither motor; and, instead, make an RLG. But if one wants to build a very sensitive (and therefore large) gyro, the PARR is preferable to the RLG. In a large-perimeter RLG, the resonator free spectral range (FSR = c/p) is small, and it is difficult to design for single longitudinal mode operation (see Figure 13.2). Therefore, if you need a laser gyro 10 m on a side, perhaps for geophysical observations, you might want to do as Shaw and others did at the Air Force Academy in Colorado Springs, and consider a PARR [5,6]. Unfortunately, technical and funding difficulties prevented them from completing their project.

The PARR Fundamental Limit

In the same way that we were able to express fundamental limits for the IFOG and the RLG, we can write a theoretical threshold value for a passive resonator [2,3,7]:

$$\Delta\Omega = \frac{\Gamma}{K(n_{ph}\eta t)^{1/2}} \tag{14.1}$$

where
   K = scale factor
   Γ = resonator linewidth
   n_{ph} = number of photons/s at the detector
   η = detector quantum efficiency
   t = sample time

We can substitute for Γ [from Equation (11.10)] with $p = N\pi d$, N being the number of turns in the resonator path (for the discrete component resonator, $N = 1$). We can use Equation (11.13) for K and Equation (12.3) for n_{ph}, and we have the alternative expression:

$$\Delta\Omega = \frac{c\lambda}{\pi d^2 NF} \left(\frac{e}{P_d Rt}\right)^{1/2} \tag{14.2}$$

These equations assume that the passive resonator has perfect servos. The detector efficiency appears because the servos can only track the resonator peaks once the detectors have observed changes in intensity greater than the electronics noise.

The Resonant Fiber-Optic Gyro

In his basic PARR patent, Ezekiel described how the resonator could be made from optical fiber, using guided wave optics rather than free-space optics; Ezekiel's group demonstrated a resonant fiber-optic gyro (RFOG) in 1983 [8]. It is much easier to control the transverse modes in a fiber waveguide by simply buying single-mode, polarization maintaining fiber (as in the IFOG). The disadvantage is that there is now a material in the resonator path, with its attendant absorption, scattering, refractive index nonlinearity (Kerr effect), and temperature sensitivity.

   Figure 14.2 shows a typical RFOG configuration; the fiber resonator may be coiled but its fiber is very short compared with the length in an IFOG coil (10 m

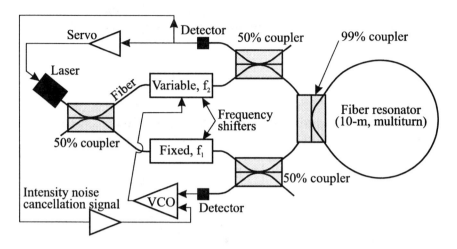

Figure 14.2. Basic resonant FOG.

instead of 300–1000 m). We saw in Chapter 12 that in the IFOG the light circulates around a coiled path of N turns, whereas in a single-turn resonator it travels around a short path N times, where N is the resonator finesse. The effect of resonance is to multiply the optical perimeter by the finesse.

In Figure 14.2 the light from the laser splits at the first 50% coupler and passes through two frequency shifters. Each path then passes through another 50% coupler and then to the resonator coupler, which transfers only 1 to 2% of the incident energy to the fiber ring. Light not passing into the ring passes through the next 50% coupler and goes to the detector. Whereas the PARR used the transmitted signals, this RFOG uses the reflected signals. Energy passing back through the frequency shifter is again frequency shifted and (half of it) reaches the laser; if it were not frequency shifted, it would interfere with the laser's mode stability and linewidth.

We mentioned in Chapter 11 that there is an optimum figure of merit [F_m, Equation (11.14)] for resonator gyros, occurring for an optimum coupling between the input waveguide and the resonator. When the RFOG ring is well off-resonance the detector signal is at a maximum, whereas at resonance all the input energy passes into the ring and none passes to the detector, in the ideal case when the coupler has this optimum value. This value is akin to the impedance matching criterion for electrical circuits and offers best performance when the coupling K_r is equal to the resonator roundtrip loss:

$$K_r = \alpha p$$

where
α = waveguide loss, cm^{-1}
p = resonator perimeter, cm

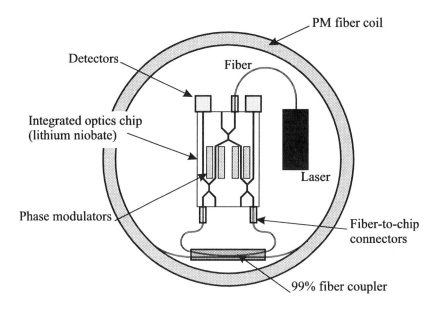

Figure 14.3. RFOG with integrated optics chip.

So for low-loss waveguides, the coupling must be very small.

Unlike the RLG and discrete PARR, the RFOG resonator can have more than one turn. More turns do not change the scale factor, but rather they give lower rate resolution, as expressed in Equation (14.2).

Whereas some RFOGs use Bragg cells, others use the serrodyne frequency shifter described in Chapter 12 [9]. Serrodyne shifters use less power but have some unique electronics complications and do not spatially separate the shifted light from the unshifted as a Bragg cell does. Figure 14.3 shows an RFOG that uses an integrated optics chip for the serrodyne frequency shifters. Attaching the resonator fibers and pigtailing the laser and detector to the chip are problems similar to those encountered in the IFOG, and just as in the IFOG, the RFOG is adversely affected by backscatter [10, 11].

The rate resolution is lower for higher detector power, and although laser power can be increased up to a point, the waveguides are so small that power density in the guide can get very large. The resonator power at resonance can reach 0.5 MW/cm^2, which can cause the third-order nonlinear terms in the refractive index to become significant—the Kerr effect [12]. While the IFOG also suffers from Kerr biases (Chapter 12), the power in the Interferometric FOG fiber is relatively low, while that in the Resonant FOG is relatively high, essentially proportional to finesse. Thus, whereas the gyro sensitivities can be made identical by making $N_{IFOG} - FN_{RFOG}$, the third order nature of the Kerr effect means that in the RFOG, a given power imbalance between the oppositely traveling waves can cause a higher gyro bias. For higher performance one may need to control the stability of the symmetry of power splitting in the waveguides, particularly against temperature.

Whereas the IFOG needs a broad band light source (with low coherence length) to avoid Rayleigh backscatter noise, the RFOG needs a high coherence, narrow linewidth laser. The laser linewidth must be narrower than that of the resonator, because the effective gyro linewidth is the convolution of the laser linewidth Γ_L and the resonator linewidth Γ, which can be expressed as a reduced finesse F' [13]:

$$F' = F/(1 + \Gamma_L/\Gamma)$$

The rate resolution equations given earlier assume that the laser line is much narrower than the resonator line, so that the effective gyro linewidth $\Gamma' = \Gamma + \Gamma_L$ and F' = F. Let us calculate what laser linewidth we need for 0.1-deg/h performance. Assume an RFOG with these parameters:

p = 10 m
d = 10 cm (N approx. 30)
n = 1.5
λ = 1.3 µm
P_d = 100 µW
R = 0.5 A/W
t = 1 s

Its pulse weight, $n\lambda/d$, is 2×10^{-5} rad/pulse or 4 arc-sec/pulse, and K = 0.25 pulses/arc-sec (or Hz/deg/h).

Rearranging Equation (14.1), and substituting for $n_{ph}\eta$:

$$\Gamma = \Delta\Omega K(P_d R/e)^{1/2} = 442 \text{ kHz}$$

This is the resonator linewidth; the smallest solid-state laser linewidths available are about 100 kHz, so that F'= 0.8F, a 20% reduction that must be compensated.

GaAlAs laser diodes ($\lambda = 0.82$ µm) costing less than $100 have linewidths over 10 MHz [14]; to get below 10 MHz at 0.82 µm requires more expensive quantum well lasers. Narrow line lasers at longer wavelengths (where fiber losses are less) are less well developed [15], and to reach 100 kHz laser builders are using external cavity designs. These currently cost more than $10,000 each, although there is no reason that they should not be cheaper, given a market pull.

Evidently, the RFOG would be expensive because it needs a special laser. Accordingly, it must be capable of high performance. In the case where one only needs low performance, at a low cost, one can use this same concept of guided-wave passive resonator gyro, substituting a waveguide ring resonator for the fiber coil resonator, and taking advantage of cheaper mass-produced diode lasers. This is the idea behind the micro-optic gyro.

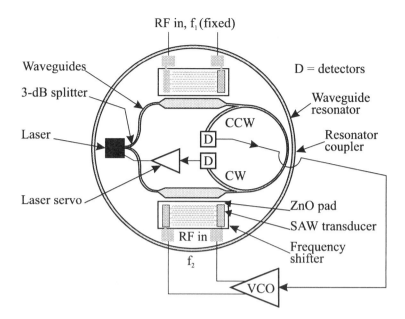

Figure 14.4. The micro-optic gyro.

The Micro-Optic Gyro

Before beginning to describe this gyro, the author must declare his lack of objectivity regarding it. Development stopped at Northrop in 1990, but the concept seems too useful to forget, and it may well be in development (perhaps in a different form) somewhere. In the description that follows we will continue to use the present tense.

The Northrop micro-optic gyro (MOG), patented in 1982 [16], is made from integrated optics, is inherently inexpensive, and uses readily available components, like the lasers mentioned that are used in CD players.

The major difference between the MOG and the RFOG is that MOG uses integrated optic waveguides instead of fibers. Fibers have very low losses, of the order of 1 dB/km (10^{-5} dB/cm), and can be coiled into small-diameter resonators of low FSR and narrow linewidth, which should give high performing gyros. Aiming at a lower performance market, the MOG uses integrated optic guides, whose losses are about 0.05 dB/cm but are inexpensive to make. MOG is aimed at the 1- to 100-deg/h market, where it should sell in production for less than $500. It could be invaluable for precision-guided munitions and GPS-aided navigators. The following analysis will show that this is a realistic aim.

Figure 14.4 shows the MOG in schematic form. Light from the laser divides and passes through two surface acoustic wave (SAW) frequency shifters, which each add a controllable frequency to the laser frequency. Light couples into the resonator in the CW and CCW directions through the directional coupler. As in the RFOG, when the ring resonates, the light passes into the resonator rather than on to the detectors, so the detectors see a minimum signal when the ring is in resonance. Rings are typically 25–50 mm in diameter, and have, naturally, a single turn.

A servo tunes the laser to maintain resonance in the CW path around the ring [17], that path having a fixed bias frequency fed to its frequency shifter, f_1. Resonance in the CCW direction comes from tuning its frequency shifter with the output from a servo-controlled voltage-controlled oscillator frequency f_2. The difference frequency $f_1 - f_2$ is the desired measure of rotation rate. The laser is butt-coupled to the input port, and chip silicon detectors are surface mounted on the output waveguides. The resonator and frequency shifters are fabricated by planar processes like those used to make integrated circuits and can use much of the same equipment.

The waveguides are regions of increased refractive index in a glass substrate. The waveguide patterns are defined photolithographically by printing a mask pattern onto a metal film coating the glass substrate. The waveguide patterns are etched through the metal, exposing the glass where the guides are to be made. The waveguides are then formed by ion exchange, a process in which the alkali metal ions (sodium or potassium) in the glass are exchanged for silver ions in a molten salt solution. This creates channels about 3 x 2 µm in section along which the light travels [18, 19].

Strong polarization maintenance is a very important benefit from fabricating waveguides in situ, rather than using a fiber. Integrated waveguides can easily be inherently linear polarized structures, which decouple the TE and TM modes, by controlling the guide width and depth. Integrated gyros also do not need the fabrication of a polarization maintaining coupler, as their couplers are formed at the same time as the resonator, in a single photolithographic step.

A crystalline ZnO piezoelectric film for the SAW frequency shifter is sputtered onto the substrate locally where the SAW will be generated. Then aluminum transducer fingers 4 µm wide are deposited on the film, which is excited by an RF signal at about 200 MHz to create the traveling SAW wave. Alternatively one can use serrodyne frequency shifters to tune a MOG [20] or use a time-division detection scheme by switching the laser alternately to the CW and CCW resonances [21]. NTT has demonstrated the feasibility of the last scheme [22] in a gyro using doped silica-on-silicon waveguides.

The simplicity and degree of integration of the MOG make it rugged. It is also an inexpensive design, because it is inherently mass-producible and will be fabricated on an automated production line like those being used for electronic IC manufacture. Using 15-cm (6-in.) diameter substrates, up to 10 gyros can be made at a time, as shown in Figure 14.5 [23].

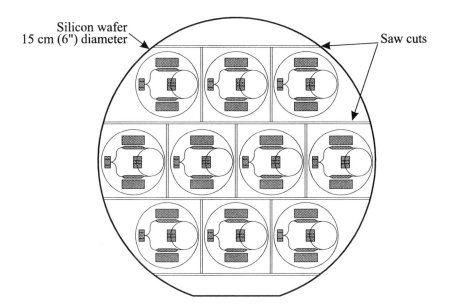

Figure 14.5. Ten MOGs on a wafer.

The MOG Fundamental Limit

The MOG shot noise limit follows from Equation (14.1), substituting for the scale factor, the laser power, the power losses in the integrated optics components, and the detector quantum efficiency. Then the MOG's minimum detectable rotation rate at the optimum coupling is

$$\Delta\Omega = \frac{C\alpha p}{K(P_L T_{chip} Rt)^{1/2}}$$

where
$T_{chip} - 10^{-(Loss/10)}$
Loss $= L_{LC} + L_{FS} + 3n_s$ dB
n_s = number of 3-dB splitters in the light path
L_{LC} = laser-to-chip coupling loss, dB
L_{FS} = frequency-shifter insertion loss, dB
P_L = laser output power
C = a constant

For best performance (lowest rate resolution) the gyro needs the lowest possible waveguide loss α, the largest possible diameter, and as much laser power as possible. Laser-to-chip coupling (L_{LC}) and frequency-shifter insertion losses (L_{FS})

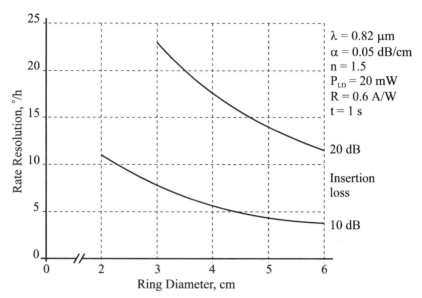

Figure 14.6. MOG rate resolution.

reduce the effective laser power at the resonator. Figure 14.6 shows some typical values, giving the performance possible for gyros with different diameter resonators. It is almost always necessary to make the gyro as small as possible to keep the guidance system small.

Waveguide loss is caused by scattering and absorption in the material and by radiation from the guides if the bend radii are small or the guide edges too rough. The MOG needs low-loss waveguides and currently integrated optics waveguides have losses of 0.05 to 0.1 dB/cm; they could beneficially be reduced to 0.001 dB/cm. But reducing losses to fiber losses, 10^{-5} dB/cm, has little advantage, since once loss reaches 0.001 dB/cm, Kerr effect and backscattering dominate the performance [24]. Backscattered light captured in the waveguide gives rise to gyro scale factor nonlinearity, whose magnitude is roughly equal to twice the backscatter for small rotation rates and small backscatter. That is, for scale factor linearity of 100 ppm, backscatter must be less than 50 ppm.

Just as we calculated for the RFOG, MOG's laser linewidth must be less than the resonator linewidth. A gyro with a 5-cm diameter resonator, with a waveguide loss of 0.05 dB/cm, will have a resonator linewidth of 37 MHz. Laser linewidth decreases with increasing power; one can easily buy GaAlAs lasers with linewidth less than 20 MHz at 30 mW nominal output power and (for ten times the price) quantum well lasers with less than 10 MHz linewidth at 100 mW.

The field of integrated optics is still maturing, being stimulated by the communications industry, which sees it as a way of making communications devices faster, cheaper, and more reliable. It is the photonic analog to the electronic integrated circuit and is moving toward the integration of devices on gallium arsenide and silicon. There is work in progress on growing GaAs devices like lasers on silicon wafers, and electronic processing might be incorporated into the same chip. Semiconductor laser development is proceeding across a broad front, from 1.5-µm GaAsInP devices for telecommunications, to GaInP 0.67-µm lasers to replace the HeNe gas laser–a vast and profitable market opportunity.

All these technology advances are being pursued for large markets other than gyros, yet all benefit MOG. Higher power lasers improve drift performance, GaAs phase modulators could reduce gyro power if used to replace the Bragg frequency shifter, and GaAs grown on silicon could lead to even less-expensive, entirely integrated MOGs with the very highest of shock and vibration tolerance. Perhaps GaAs laser-phase modulator chips will be coupled to waveguides grown on silicon wafers (glass, after all, is mostly silicon dioxide).

IFOG, RFOG, and MOG Size Limits

There is really no limit to the largest size that one can make an IFOG or RFOG for operation in a benign environment. One could conceivably coil a fiber entirely around a space satellite if necessary or put a large coil around the complete INS in a ship. The limit to the smallest size is the loss due to bending of the fiber, because with finite losses one cannot compensate for small coil area with more turns, or with higher finesse. Fiber gyros need a coil former and some connectors from the coil to the integrated optics chip, all of which must be packaged, but this is reasonably straightforward engineering. Because of its shorter fiber coil, an RFOG would be a little smaller than an IFOG.

It is unlikely that MOGs can be usefully made larger than 10 cm in diameter, because they need a substrate to support the resonator. They can more easily be made small because they have no structural "overhead" like coil formers and fiber connectors, because the sensing resonator is integrated. Bending losses set the low limit to resonator ring diameter at about 10 mm, but because one would want to package the shifters inside the ring (for greatest gyro sensitivity in a given size), in practice the frequency shifter size keeps the resonator diameter to at least 35 mm.

Fundamental Limits for RFOG, IFOG, and RLG

Now that we have studied the three important optical gyros, we are in a position to compare their fundamental limits. Following Ezekiel and Arditty [7], we can write their fundamental limits as follows:

IFOG shot noise limit, Equation (12.2):

$$\Delta\Omega = \frac{c\lambda}{2Ld(n_{ph}\eta t)^{1/2}} = \frac{c\lambda}{2\pi d^2 N(n_{ph}\eta t)^{1/2}} \tag{14.3}$$

RLG, spontaneous emission noise, Equation (13.2):

$$\Delta\Omega = \frac{\lambda p \Gamma}{4A(n_c t)^{1/2}}$$

where Γ = cold cavity linewidth, without gain medium.
Approximately, for a circular cavity:

$$\Delta\Omega = \frac{c\lambda}{\pi d^2 F(n_c t)^{1/2}} \tag{14.4}$$

RFOG (PARR), shot noise limit, Equation 14.1:

$$\Delta\Omega = \frac{\lambda p \Gamma}{4A(n_{ph}\eta t)^{1/2}} = \frac{c\lambda}{\pi d^2 NF(n_{ph}\eta t)^{1/2}} \tag{14.5}$$

Let us assume, reasonably:

1. gyros with the same d, λ, and t;
2. comparable optical power ($n_c = n_{ph}$);
3. η approximately = 1 ;
4. we can ignore the factor 2 in the denominator of Equation (14.3); and
5. that $N_{IFOG} = F_{RLG} = (NF)_{RFOG}$.

Then we can compare Equations (14.3)–(14.5) and see that the three types of gyro have theoretically much the same random drift performance.

In practice the RLG usually has a much higher finesse than the PARR; RLGs can have finesses around 10,000, whereas passive gyros are hard pressed to reach 100. For an RFOG to match RLG performance it would need to have 100 turns and still achieve this finesse. And for an IFOG to have this same performance it would need to have a coil of 10,000 turns, 10 times the usual number.

The different wavelengths used in practical gyros affect these comparisons to a degree. RLGs usually use the visible HeNe transition at 0.63 μm, whereas many fiber gyros use 1.3- or 1.55-μm semiconductor sources and, accordingly, all else being equal, have double the theoretical resolution. MOG used the 0.82-μm semiconductor laser. In passing, observe that the RFOG's pulse weight is greater than that of the same-size RLG, caused by the larger index of the fiber light path and the longer wavelength.

Conclusion

It is important to point out again that no passive resonator gyros have yet, in 1998, been developed and put into production. They do hold some promise for particular applications, which is why they have their own chapter.

A major difference between the IFOG and the RFOG is that the RFOG is more susceptible to Kerr effect bias errors [25, 26]. The resonant power increase that we mentioned earlier is exacerbated by the spectral density of the RFOG's narrow line laser.

The theoretical performance limits tell only part of the story of a successful product. The RLG cavity will always need to be precisely fabricated, will need very low-scatter mirrors, and demands absolutely tight seals between the cavity and the mirrors and electrodes. Consequently, RLGs are expensive; even so, their reliable performance should continue to make them useful for many years.

The attachment of a FOG's fiber to its processing chip needs a mechanical joint, not easily made rugged and stable. Fibers and light sources for FOGs are special and thus expensive, and while it is true that the fiber gyros should have better drift performance than MOG, there are fields that do not need higher performance. They need low cost, small size, and reliability, all best achieved by total integration.

To sum up the optical gyros: for the best performance, use an RLG. For day-to-day drift repeatability of the order of 0.1–0.001 deg/h and high reliability, use an IFOG. For lowest cost and extremely high shock tolerance (e.g., in smart cannon-launched munitions) develop a MOG!

References

1. Ezekiel, S., "Laser gyroscope," U.S. Patent 4 135 822, 23 Jan. 1979.
2. Ezekiel, S., S.R. Balsamo, "Passive ring resonator laser gyroscope," Appl. Phys. Lett., 30, 9, pp. 478–480, 1 May 1977.
3. Sanders, G.A., M.G. Prentiss, S. Ezekiel, "Passive ring resonator method for sensitive inertial rotation measurements in geophysics and relativity," Opt. Lett., 6, 11, pp. 569–571, Nov. 1981.
4. Zarinetchi, F., S. Ezekiel, "Observation of lock-in behavior in a passive resonator gyroscope," Opt. Lett., 11, 6, pp. 401–403, June 1986.
5. Shaw, G.L., B.J. Simmons, "A 58 m² passive resonant ring laser gyroscope," in Moore, E.L., O.G. Ramer, (Eds.) *Fiber-Optic and Laser Sensors II*, Proc. SPIE 478, pp. 117–121, 1984.
6. Rotgé, J.R., G.L. Shaw, B.J. Simmons, "Rotation sensing at 10^{-10} earth's rate," DGON Symposium Gyro Technology, Stuttgart, 1985.
7. Ezekiel, S., H.J. Arditty (Eds.) *Fiber-Optic Rotation Sensors*, Springer-Verlag, New York, 1982.
8. Meyer, R.E., S. Ezekiel, D.W. Stowe, V.J. Tekippe, "Passive fiber-optic ring resonator for rotation sensing," Opt. Lett., 8, 12, pp. 644–646, Dec. 1983.

9. Sanders, G.A., G.F. Rouse, L.K. Strandjord, N.A. Demma, K.A. Miesel, Q.Y. Chen, "Resonant fiber-optic gyro using LiNbO$_3$ integrated optics at 1.5 μm wavelength," *Fiber-Optic and Laser Sensors VI*, Proc. SPIE 985, pp. 202–210, 1988.

10. Iwatsuki, K., K. Hotate, M. Higashiguchi, "Effect of Rayleigh backscattering in an optical passive ring resonator gyro," Appl. Opt., 23, 21, pp. 3916–3924, 1 Nov. 1984.

11. Takahashi, M., S. Tai, K. Kyuma, K. Hamanaka, "Effect of reflections on the drift characteristics of a fiber-optic passive ring resonator gyro," OFS Technical Digest Series, 2, part 2, pp. 401–404, 1988.

12. Iwatsuki, K., K. Hotate, M. Higashiguchi, "Kerr effect in an optical passive ring resonator gyro," J. Lightwave Tech., LT-4, 6, pp. 645–651, June 1986.

13. Gröllmann, P., J. Herth, M. Kemmler, K. Kempf, G. Neumann, S. Oster, W. Schröder, "Passive fiber resonator gyro," DGON Symposium Gyro Technology, Stuttgart, 1986.

14. Ho, S.T, S. Ezekiel, J.R. Haavisto, J.J. Danko, "Optical feedback stabilization of a semiconductor laser," J. Lightwave Tech., LT-4, 3, p. 312, 1 Mar. 1986.

15. Ohtsu, M., S. Araki, "Using a 1.5 μm DFB InGaAsP laser in a passive ring cavity type fiber gyroscope," Appl. Opt., 26, 3, pp. 464–470, 1 Feb. 1987.

16. Lawrence, A.W., "Thin film laser gyro," U.S. Patent 4 326 803, 27 April 1982.

17. Danko, J., "Integrated optic diode laser stabilizer," *5th European Conference on Integrated Optics,* Proc. SPIE 1141, pp. 52–58, April 1989.

18. Eguchi, R.G., E.A. Maunders, I.K. Naik, "Fabrication of low-loss waveguides in BK7 by ion exchange," Proc. SPIE 408, Mar. 1983.

19. Haavisto, J., "Thin film waveguides for inertial sensors," Proc. SPIE 412, 221, 1983.

20. Segré, J., J.R. Haavisto, "Thin film passive ring resonator laser gyro," U.S. Patent 4 658 401, 14 April 1987.

21. Haavisto, J., "Resonant waveguide laser gyro with a switched source," U.S. Patent 4 661 964, 28 April 1987.

22. Iwatsuki, K., M. Saruwatari, M. Kawachi, H. Yamazaki, "Waveguide-type optical passive ring resonator gyro using time-division detection scheme," Electron. Lett., 25, 11, pp. 688–689, 25 May 1989.

23. Lawrence, A.W., "Providing an inexpensive gyro for the navigation mass market," ION National Technical Meeting, 23–25 Jan. 1990.

24. Lawrence, A.W., "The micro-optic gyro," DGON Symposium Gyro Technology, Stuttgart, 1983.

25. Carroll, R., C.D. Coccoli, D. Cardarelli, G.T. Cote, "The passive resonator fiber-optic gyro and comparison to the interferometer fiber gyro," in Udd, E. (Ed.), *Fiber-Optic Gyros: 10th Anniversary Conference,* Proc. SPIE 719, pp. 169–177, 1987.

26. Frigo, N.J., "A comparison of interferometric and resonant ring fiber-optic gyroscopes," *Fiber-Optic and Laser Sensors VI*, Proc. SPIE 985, pp. 270–277, 1988.

15
Testing Inertial Sensors

In Chapter 2 we described inertial systems error modeling and the generation of specifications for the sensors in an inertial navigator, specifications which could (and should) follow the Institution of Electrical and Electronic Engineers (IEEE) standard formats for inertial sensors. In this chapter we will describe the testing of accelerometers and gyros for conformance to these specifications; we will describe the tests that we would perform for acceptance test procedures (ATPs) on a pendulous accelerometer and on three types of gyro, the single-degree-of-freedom (SDFG), the dynamically tuned gyro (DTG), and the ring laser gyro (RLG). This description is general and takes account of neither specific design issues nor of particular customer needs that would expand the ATP in real life.

When we test accelerometers and gyros with drifts below 50 μg and 10 deg/h, respectively, we need a properly designed and equipped test laboratory, a facility outfitted with precise mechanical positioning equipment and electrical standards.

Inertial Sensor Test Labs

If we wish to verify the performance of a gyro, we need to know the input rotations to the gyro with respect to inertial space, i.e., the earth's rotation rate component about the gyro's input axis. For an aircraft navigator gyro, with an accuracy of 0.01 deg/h, we must be sure that the fixture holding the gyro is fixed to the earth to a stability of better than 10^{-4} rad (20 arc-sec), because 10^{-4} x HER (say 10 deg/h) is 0.001 deg/h, an order better than we are trying to measure. Similarly, for testing an accelerometer for 1 μg stability, the test base must be aligned to the vertical to better than 1 μrad (0.2 arc-sec).

We need to isolate the sensor from ground vibrations, like those from passing trucks or trains, therefore inertial systems manufacturers and prime users have test labs supplied with stable piers, which are massive concrete slabs set on bedrock or in deep sand, isolated from the foundations of the building containing the lab. Some test piers have been supported on air bags, giving the pier a very low natural frequency. The pier at the Air Force Academy Seiler Laboratory in Colorado Springs weighs 450,000 pounds, is floated on pneumatic isolators, and is actively controlled by precise tilt meters; it is stable to better than $10^{-8}g$ and 10^{-3} arc-sec. Probably the best known inertial test lab in the United States is the Central Inertial

Guidance Test Facility (whose acronym is pronounced "CIG-TIF") at Holloman Air Force Base, New Mexico.

The direction of True North must be precisely known for proper compensation for the earth's rate, so labs have sight lines from the test pad to the pole star (perhaps through holes in the roof). If this is not feasible, perhaps because the test pad is in a basement on bedrock, star sights are indirectly transferred to designated positions (marked by points in the floor) from outside the building, and retro-reflectors built into the wall are aligned from the star sights. One then aligns the gyro test stands to the retro-reflectors.

Performance Test Gear

Sensor test stands can be turned and tilted to position the sensor axes up and down and along the cardinal points of the compass, thereby allowing gravity and/or and the earth's rate to act along different axes in turn. These tables have arc-sec positioning accuracy and usually have precise spirit levels installed for setting horizontal accurately. They are electrically driven and provide digital outputs of trunnion angles, so they can be programmed to perform tests automatically. This provides repeatability of test conditions and times while reducing personnel costs.

Precise rate tables, controlled so that they can spin up to 1000 deg/s or so about a well-defined axis, are used for measuring the rate gyro's scale factor. The table will have slip rings to take power to the gyro and signals out to the measuring electronics and will have a precise angle marker, which gives a pulse out per revolution, although some tables provide precise rotation rate (or incremental angle pulses) continuously.

Another type of rate table, the servo table, simulates a single-axis inertial platform (Chapter 1) by having a platen mounted on precise bearings, rotated by a motor. The motor will turn the table platen when commanded by a servo loop signal, the platen speed being proportional to the magnitude of the signal and its direction depending on the signal polarity.

The servo drive signal is obtained from the pickoff of a gyro mounted on the table with its input axis (IA) along the table axis, as shown in Figure 15.1. In order to measure the torquer scale factor of a single-axis floated gyro (SDFG) or a dynamically tuned gyro (DTG), a current is sent in through the table slip rings to the gyro torquer. This causes the gimbal (SDFG) or rotor (DTG) to precess about the output axis (OA) and generates a pickoff signal. This signal passes out through the table slip rings to the external servo, which excites the table motor. The table accelerates until the gyroscopic torque exactly equals the current-induced torque and the gyro pickoff returns to null. Because it needs only a precise current source and a clock to measure table rate, such a table provides a higher accuracy measurement of scale factor than an open-loop rate table, and it is used to determine gyro torquer linearity as described later.

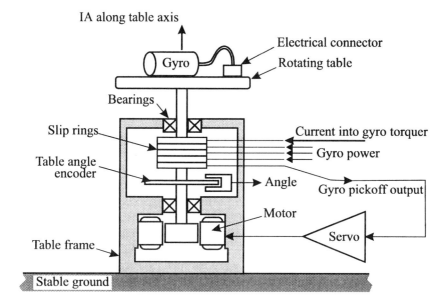

Figure 15.1. The servo table.

The test stands and rate tables might be equipped with variable temperature enclosures so that sensor parameters can be measured over temperature. The whole lab will generally be temperature controlled to a few degrees Celsius. Magnetic field coils can be mounted on a test table for checking magnetic sensitivity.

Each test table in the lab will be equipped with an electronics console containing the instrument power supplies, servo loops, and output circuits, usually of better performance than those used in the navigation system, so that one is sure that one is testing the sensor, not the electronics. The data will be recorded by a computer and processed by programs that fit the data to predetermined models, calculate residuals, and derive error reports.

Environmental Test Gear

An environmental test laboratory is equipped with vibrators capable of shaking a sensor (or a whole guidance system) with random vibrations up to 1 g^2/Hz from 20 Hz to 2000 Hz. They will be able to generate sine vibrations up to 50g. Shock pulse testers might be provided, which slam the test piece against a pad of material chosen to give the desired acceleration versus time profile. A centrifuge may be available for generating steady accelerations, and some models have counter-rotating tables at the end of the arm so that the sample rotation can be "unwound"; gyros on the test plate will thus see no net rotation, but will be exposed to an acceleration that seems to circle around the sample [1].

The CIGTF has a gigantic centrifuge, a 47-ton machine with a 260-in. (6.5-m) arm; it has a counterrotating table. It can provide from $0.25g$ to $100g$ radial acceleration, and even though the centrifugal force stretches the arm 0.0015 in/g, it repeats from revolution to revolution to better than 10 ppm. For testing under purely linear accelerations, such as those found during an ICBM's boost phase, CIGTF has a rocket-propelled sled that can carry a guidance system over a precisely surveyed track [2].

Chambers are provided that can bake and freeze sensors or systems in atmospheres of varying humidity. Some labs can expose systems to sand and fungi to simulate storage and handling in the field, and to salt spray, simulating conditions at sea. Sensors have been known to develop electrical problems when exposed to salt spray, which seeped inside the radar domes of an anti-ship guided missile.

Qualification, Acceptance, and Reliability Tests

There are three types of test commonly carried out on sensors: qualification, acceptance, and reliability tests. *Qualification tests* precede production; they are the most stringent tests done and are intended to show that a particular design will meet all the customer's requirements with ample margin for production tolerances. In the qualification tests, the manufacturer will operate a sample of sensors through the extremes of temperature, vibration, shock, magnetic field, and anything else the inertial system will experience. These tests are usually done only once, at the beginning of a program; the qualification test samples must represent the production build, and once a sensor has passed "qual tests" the design should be considered frozen (unchangeable). If the manufacturer makes a substantial change to the design or fabrication processes used in building the sensor, the manufacturer may be obliged to "re-qual," to redo the qualification tests, an expense not incurred lightly.

Then, during production, every sensor will undergo an *acceptance test procedure* (ATP), testing that checks selected parameters and determines calibration constants for system use. The customer will receive the resulting data in an agreed-upon test report. Some tests may be done on only a sample, selected using some rules agreed between the supplier and the customer. If the sample fails the test, shipments will be stopped until the cause is identified and the problem fixed. ATP often consists of bias, scale factor, random drift, and temperature tests; some vibration tests might be included as a check for good workmanship.

Sometimes an ATP will include tests peculiar to a single model sensor. For example, if a gyro has exhibited high output noise in the system, but has supposedly been cured, some test might be designed to verify that this is so. Ball bearings are prone to occasional instability (caused by the retainer whirling around inside the bearing rather than smoothly rotating), behavior depending on the amount of oil and its distribution, the preload, and the clearances, changing over temperature. If the customer felt nervous about the gyro design or the manufacturing build control, the customer might insist that every gyro had a noise test over temperature.

In the third group of tests, *reliability tests*, a sample of sensors, taken at random from the production line, will be run in simulated operating conditions to see how long they last. Reliability tests aim at deriving *mean time between failures* (MTBF; see Chapter 16). Gyro wheel bearing life tests can verify that bearing fabrication and processing remain under control; ring laser gyros will be run to ensure that helium gas seals are reliable, and DTGs may be vibration tested for hours to check that flexure processing maintains fatigue life.

Accelerometer Testing

The manufacturer will provide a group of constants that are nominally true for every gyroscope or accelerometer of a particular model; they are not necessarily checked on every one. For a pendulous accelerometer these reference constants may be the moment of inertia of the pendulum, the pendulosity, the damping coefficient, the pickoff sensitivity, the forcer scale factor (as distinct from the instrument scale factor), and the proof mass.

The Accelerometer Acceptance Test Procedure

To begin, we check the sensor for warm-up trends by looking at the output for a time after it is turned on. Then, all being well, we check the threshold and hysteresis (Chapter 2). To check hysteresis, we displace the pendulum from its null position by adding an offset voltage into the servo loop. The hysteresis is the bias remaining when the offset is removed and the servo has driven the pendulum to the operating null. Now that we are satisfied that the instrument has been properly built, we proceed to determine its calibration constants.

We referred to the IEEE model equation for a pendulous accelerometer in Chapter 4; it says that the indicated acceleration is a function of the actual accelerations along the input axis (IA), the pendulous axis (PA), and the output axis (OA):

$$E/K_1 = K_0 + a_i + K_2a_i^2 + K_3a_i^3 + d_oa_p + K_{ip}a_ia_p - d_pa_o + K_{io}a_ia_o \qquad (15.1)$$

where
   E = accelerometer output signal
   a_i, a_p, a_o = accelerations along the IA, PA, and OA
   K_0 = bias, g
   K_1 = scale factor, output units/g
   K_2 = nonlinearity, g/g^2
   K_3 = nonlinearity, g/g^3
   d_o, d_p = IA misalignment with respect to the OA and PA
   K_{ip}, K_{io} = cross-coupling coefficients, g/g^2

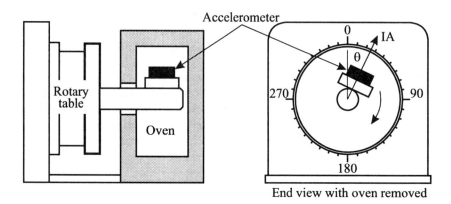

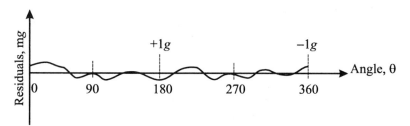

Figure 15.2. Plus/minus 1*g* tumble test.

In the ATP we measure the K and d coefficients and their variations with time and temperature. We use gravity as the test acceleration, by attaching the accelerometer to a precise rotary table and rotating it about a horizontal axis so that the sensor axes are exposed to gravity in each direction, commonly referred to as a "plus and minus 1*g*" excitation tumble. As gravity is not the same in every location, we must report the results normalized to a standard gravity value (Chapter 1).

For a rotary table we often use a machinist's dividing head because it is rigid, stable, and can set rotation angle to arc-sec accuracy. A dividing head with a stepping motor and an encoder can be computer controlled and the test procedure automated. The head axis can be set horizontal using a precise right-angle block and a level. As shown in Figure 15.2, the head may be outside a combination oven and refrigerator (e.g., a Tenney chamber), with an axle extending into the chamber through a hole in its side, and the units under test are mounted at the end of the axle.

Assuming that the accelerometer gives an analog output with scale factor K_1 V/g, the output at any angle θ is

$$V/K_1 = g \cos \theta + K_0$$

Fitting the output to a best cosine model, the scale factor K_1 and bias K_0 are deduced. They are then used to calculate the residuals (the differences between the actual outputs and those predicted by the calculated scale factor and bias), which

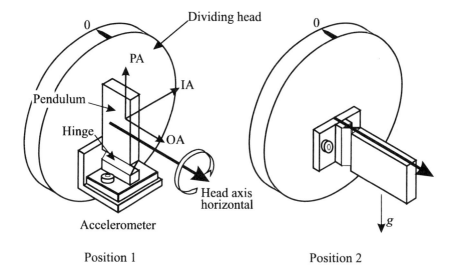

Figure 15.3. Accelerometer test positions.

may be as sketched in Figure 15.2. There will be some specified limit on the standard error, that is, the r.m.s. of the residuals.

The accelerometer can be mounted in either of two positions shown in Figure 15.3, depending on which error terms are to be measured. The test fixture that holds the accelerometer should be rigid so that it does not deflect as it is rotated (this will appear as an axis misalignment), and it should approximate the same thermal and electrical properties as the system in which the accelerometer will be used.

First we set the accelerometer in position 1, with its OA parallel to the (horizontal) head axis, its PA up, and its IA horizontal, and take output measurements E_0, E_{90}, E_{180}, and E_{270} at $0°$, $90°$, $180°$, and $270°$ angles around the head. When the accelerometer IA points up and down ($90°$ and $270°$) it sees $\pm 1g$, so we can calculate the scale factor:

$$K_1 = \frac{1}{2}(E_{90} - E_{270}) \text{ units}/g$$

We can calculate the bias, in units of g, from the sum of these readings:

$$K_o = \frac{1}{2}(E_{90} + E_{270})/K_1$$

When the accelerometer IA points horizontally ($0°$ and $180°$) it should see no gravity component, so the bias will also be

$$K_o = \frac{1}{2}(E_0 + E_{180})/K_1$$

These two readings also give us the IA misalignment with respect to the OA:

$$d_o = \frac{1}{2}(E_0 - E_{180})/K_1 \text{ rad}$$

Mounting the accelerometer in the second position, with the PA horizontal along the head axis, the OA down, and the IA horizontal, allows one to calculate the IA misalignment to the PA:

$$d_p = \frac{1}{2}(E_{180} - E_0)/K_1 \text{ rad}$$

These multiposition tests are done many times, and the best estimates of the model coefficients are found statistically. In an ATP we repeat these tests at high and low temperatures, perhaps at $-50°C$ and $+80°C$, depending on the customer's specification. We will also test for the long-term stability of each of the coefficients (the *turn-on to turn-on* repeatability), by testing the sensor and then exposing it (not operating) to shocks and temperatures before retesting it. The environments chosen might simulate handling from the factory to the user or other nonoperating environments.

We test for in-run bias stability by putting the sensor on a stable base (colloquially known as a *tombstone*) and sampling the output over a time that might represent one mission. This might be a few minutes for a tactical missile, hours for an airplane, or days for a submarine guidance sensor. The standard deviation of the output about the mean will be a measure of the accelerometer's random drift and will have a specified maximum limit.

Centrifuge Tests.
As the multiposition tests are done only at one acceleration ($\pm 1g$) we cannot separate the higher-order terms in Equation (15.1), and K_o will include K_2, whereas K_1 will include K_3; neither can we separate misalignment from cross-axis sensitivity. To find these coefficients we must use a centrifuge. As the radial acceleration is $r\omega^2/g$ (where r = arm length, or radius to the test point, and ω = angular rotation speed), we must know the rotation speed and arm length accurately and allow for the arm stretching with increasing speed. Then we can measure the accelerometer output as a function of acceleration along the principal axes and half-way between them, and fully determine the model coefficients, including scale factor linearity and asymmetry. We can also check to ensure that the instrument can reach the full-scale value for which it was designed. Because a precise centrifuge is expensive, we would not usually use one for an ATP, although Peters [3] has described a procedure for using reference accelerometers to increase the accuracy of measurements made with an inexpensive centrifuge. The IEEE is preparing a standard procedure for accelerometer centrifuge testing [4].

Although these general test procedures make a good starting point, test specifications are always tailored to the system design and the mission requirements. They also include tests specific to the sensor design; testing a vibrating beam accelerometer (VBA, Chapter 5), for example, might include a check of the vibropendulous error around the pendulum resonance.

Gyroscope Testing

In this section we will describe test procedures for rate, rate integrating, tuned, and laser gyros.

Testing the SDF Rate Gyro

For the rate gyro, which performs in the 100-deg/h domain, we measure the scale factor, composite error, hysteresis, zero offset (bias), threshold and resolution, and *g*-sensitivity (Chapter 2). The rate gyro's lower accuracy means that the test equipment is relatively unsophisticated; a rate table will be used for most measurements.

After basic checks of resistance and insulation strength, polarity, and wheel run-up and run-down, the gyro is run on the rate table at one or more temperatures. With the gyro's input axis (IA) along the table axis (which is usually vertical), the table rate is varied in steps and the gyro output noted. The rates will sequence from maximum positive, to zero, to maximum negative, to zero, to maximum positive, and so on for a few cycles. The table must accelerate smoothly from speed to speed and must not overshoot a reading. The output might plot to something like Figure 2.1; the scale factor (K) will be the best-fit straight line to the data, the composite error will be the maximum deviation from that scale factor line, and the offset will be the intercept of the best-fit line on the y-axis. The hysteresis will be the difference between the two values (on the ascending and descending arms of the loop) at zero input, as shown in Figure 2.2. All the readings are scaled to rate using the calculated scale factor.

To find the threshold and resolution, gradually increase the table rate until the gyro output is half what you would expect the change to be, given the scale factor. That rate is the required value; the difference between threshold and resolution is that the threshold detects small changes from zero rate, whereas resolution detects small changes from some steady input rate.

To measure the *g*-sensitive coefficients (mass unbalances) along the IA and the spin axis (SA), place the gyro with its output axis (OA) horizontal, pointing North. Rotate the gyro about the OA in four steps so that the IA is, in turn, up, East, down, and West, measuring outputs E_U, E_E, E_D, and E_W, respectively. The SA and IA mass unbalances are

$$D_S = \tfrac{1}{2}(E_E + E_W)/K$$
$$D_I = \tfrac{1}{2}(E_U + E_D)/K$$

The detailed steps for carrying out these tests are given in the IEEE standard [5].

Testing SDF Rate-Integrating Gyros

Because SDFGs have been around for so long, their test procedures are well developed, and the IEEE standard [6] thoroughly describes them. The tests are designed to measure the calibration constants such as command rate scale factor, gyro gain, and characteristic time, and to measure the coefficients of the model equation, such as bias and mass unbalances. They also check gyro health, looking for gimbal friction (stiction), unusual bearing noises, run-up and run-down times, and so on.

As for the rate gyro, one begins by checking electric circuits for continuity, resistance, and insulation resistance and then checks spin motor phasing to be sure that the wheel turns in the right direction. Gas bearing wheels generally operate in only one direction because of the pumping grooves they carry (Chapter 7), and quite a few have been ruined by improper wheel supply phasing. Then one measures starting power, running power, run-up time, and run-down time. Measurements of the variation in wheel power at the milliwatt level can tell one what is going on inside the ball bearing wheel, because changes in lubrication or in bearing stability are accompanied by small power changes.

A useful test checks float (gimbal) stiction using a displacement hysteresis test. With the pickoff and torquer connected in a low gain servo, the float is displaced about the OA to each stop in turn by offsetting the pickoff signal. Upon removing the offset, the float should return to pickoff null; the difference in torquer currents after return from each stop represents the displacement hysteresis. Stiction may be due to misflotation, poor OA suspension quality, and/or dirt in the fluid.

Measuring the open-loop scale factor, the gain, and the characteristic time (the *transfer characteristics*) completes the preliminary checks. The open-loop scale factor relates the pickoff output to the rotation angle about the IA and is the gyro gain (H/c, Chapter 7) multiplied by the pickoff sensitivity (called "pickoff scale factor" in the IEEE standard). Note that we have generally been referring to closed-loop scale factor in this book; the open-loop scale factor K_o is measured by rotating the operating gyro a known angle about the IA and noting the pickoff output. The pickoff sensitivity K_{po} is known from a preassembly test, so that the gyro gain is given by K_o/K_{po}.

We are now ready to determine the drift coefficients for this gyro, defined in the IEEE model equation (Chapter 7). It has the steady-state form:

$S/K = \Omega$ Inertial rate about the IA

$+\ B$ Bias

$+\ D_I a_I + D_O a_O + D_S a_S$ g-sensitive drifts

$+\ D_{II} a_I^2$ g^2-sensitive drifts

$+\ D_{SS} a_S^2$

$+ D_{IS}a_I a_S$

$+ D_{IO}a_I a_O$

$+ D_{OS}a_O a_S$

where S is the output signal (perhaps torquer current required to hold the gimbal at null), K is the gyro scale factor, and all other terms are as defined in Chapter 7. For platforms, we only need to know the scale factor at low rates, because we command the platform (through the gyro torquer) to align it at only 100 times earth rate (0.5 deg/s) or so. We will carry out:

1. tombstone tests to determine random drift,
2. six-position tests to find scale factor and some drift coefficients,
3. equatorial tumble (polar-axis) tests to look at stiction and acceleration-squared coefficients, and
4. vibration tests for a more thorough determination of g^2 terms.

We will now describe these tests in more detail, but for a really thorough (working-level) study, you should refer to the IEEE standard.

Tombstone Tests.
The gyro is mounted in its test fixture on a stable base and operated closed-loop, with the OA horizontal, and the IA up. It is allowed to run for some hours, sampling the output. For gyros down to 0.01 deg/h, the random drift is determined by sampling the gyro output (current if in analog loop, pulses if in digital loop) over periods of 10–100 s until some hundreds of samples had been recorded. For higher-performance gyros, such as those used in submarine navigation, the sample times are longer, and the drift may be sampled for days, simulating long submerged voyages.

Looking at the drift data, the tester would see if the mean drift over a few minutes was wandering erratically, or if the drift tended to vary in the same direction with time (a ramp). Sometimes the ramp (or a correlated periodic function) is removed before proceeding. The random drift is the standard deviation of the samples about the mean and must be within specified limits for the gyro to be acceptable. Noisy bearings, a broken output axis suspension, kinked flexleads, and contaminated flotation fluid are some problems that cause a gyro to fail this test.   The mechanical gyro's random drift follows no fundamental law, although it usually falls with increasing sample time. The frequency spectrum of an SDFG or a DTG may have peaks at the pickoff excitation frequency, the motor excitation, the wheel speed, bearing frequencies, and beats between all these. For consistent data one needs to specify the test sample time and the servo loop type, gains, and bandwidth.

The Six-Position Test.
In this test one finds the low-rate torquer scale factor, the bias, and the mass unbalances along the IA and SA. Some testers regard this as two separate tests,

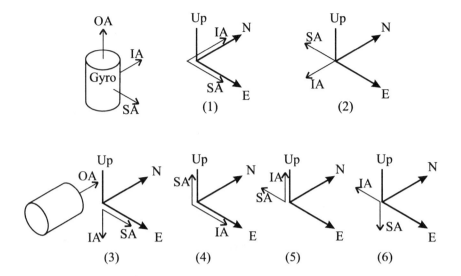

Figure 15.4. Six-position test sequence.

calling the second part a "four-point balance check." Figure 15.4 shows the six positions; 1 and 2 have the OA vertical and the IA North and South, while 3–6 have the OA horizontal. We get acceleration-sensitive coefficients from local gravity and torquer scale factor from horizontal earth's rate, so we must know both λ and g at the test lab. The horizontal and vertical components of the earth's rate are, respectively,

$$\Omega_{eh} = \Omega_e \cos \lambda$$
$$\Omega_{ev} = \Omega_e \sin \lambda$$

We assume for simplicity that the gyro axes are accurately aligned to the table axes, which are accurately aligned vertically and North. If they are not, we need to include misalignment angles in the analysis. Recording the gyro output x_n at the nth position, we get:

Position 1: $Kx_1 = B + \Omega_{eh}$

Position 2: $Kx_2 = B - \Omega_{eh}$

Position 3: $Kx_3 = B + D_I - \Omega_{ev}$

Position 4: $Kx_4 = B - D_S$

Position 5: $Kx_5 = B - D_I + \Omega_{ev}$

Position 6: $Kx_6 = B + D_S$

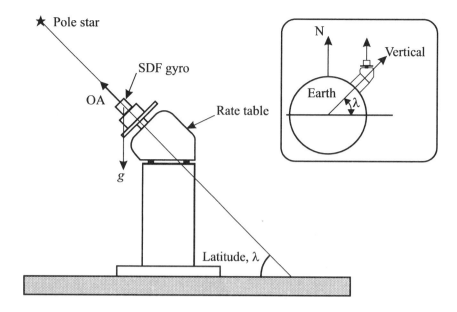

Figure 15.5. Polar-axis tumble test.

Solving these equations in groups we get:

$$K = 2\Omega_{eh}/(x_1 - x_2)$$

$$B = \tfrac{1}{2}K(x_1 + x_2) \qquad \text{from the OAV positions}$$

$$D_S = \tfrac{1}{2}K(x_6 - x_4) \qquad \text{from the OAH positions}$$

$$D_I = \tfrac{1}{2}K(x_3 - x_5)$$

We can calculate another bias value from the 3–6 position data and compare it with the OAV data. If the two do not agree, we have a gyro with a sensitivity to acceleration along the OA, D_O in the model.

The Polar-Axis (Equatorial Tumble) Test.
Although not often used these days, the polar-axis tumble test is a useful qualitative screening test for SDFGs. One mounts the gyro on a rate table with its OA accurately aligned parallel to the table axis, because then the gyro does not sense any component of the table rate [7]. Figure 15.5 shows that the rate table is tilted at the latitude angle so that its axis is parallel to the earth's rotation axis, and the gyro does not sense earth's rotation. But as the table rotates, gravity acts on the gyro, so this test excites acceleration-sensitive errors like mass unbalances, aniso-elasticity, bubbles in the flotation fluid, and so on.

The gyro is operated closed-loop and the output signal plotted as the table turns at about 150 deg/h. The gyro bias offsets the output signal, and the mass unbalance is the half-amplitude of the fundamental sine wave at table rotation frequency. The second harmonic is the anisoelasticity. Bubbles and dirt in the fluid will appear as steps in the data as the gimbal hangs up and a larger torque has to be applied to overcome the obstacle.

Some testers like to halt the table rotation for a while when the IA points in the four cardinal directions and do a displacement hysteresis test. The speed and smoothness by which the gimbal returns to null show the performance of the flexleads and the OA bearings.

The Servo Table Scale Factor Test.
The earth's rate scale factor determined earlier does not provide sufficient information for gyros used in strapdown systems. We need to know linearity and asymmetry over the operating range (perhaps ±400 deg/s), with an accuracy better than 10 ppm.

The gyro is mounted with its IA exactly collinear with the axis of a servo table whose axis is vertical, as in Figure 15.1. For closed-loop gyros (SDFG, DTG) we use the procedure mentioned at the beginning of this chapter, injecting a torquer current into the gyro and servo-driving the table to maintain pickoff null. If the rates are high, care must be taken that the torquer does not overheat.

For open loop gyros (e.g., ring laser gyro and fiber-optic gyro), the table is driven at different rates, measured by timing whole revolutions. The gyro output is measured over the same interval; for digital loops, a counter is gated by the table revolution marker pulse. The output is measured for rates covering the user's needs.

The data are fitted to an agreed model, perhaps as simple as a straight line, as in Figure 15.6. In that case a best-straight-line fit and the residuals (the differences between the true rates and the outputs scaled by the estimated scale factor) are calculated, and either the standard error of the residuals or the maximum deviation from the line is specified. Asymmetry might be modeled by different best-fit lines for positive and negative rates, scale factors K^+, K^-. Higher-order models can be used; the degree of modeling possible depends on the repeatability of the gyro coefficients. But complex models use more computer memory and bite into the computation cycle time available for strapdown algorithm processing; quadratic models are often used.

Vibration Tests.
The polar-axis test does not give an accurate figure for the g^2 drift coefficient because the applied acceleration is so low; it varies from $1g$ at the equator (where the test table would be horizontal) to zero at the poles. Electromechanical vibrators can apply $10g$ or more to the gyro and are thus better for measuring g^2 terms. But they are notorious for leaking a magnetic field big enough to upset a gyro, and one must either shield the gyro or use compensating field coils.

The gyro is mounted on a vibrator, taking care that the vibrator axis moves precisely linearly without imparting any rates to the gyro. One way of ensuring this

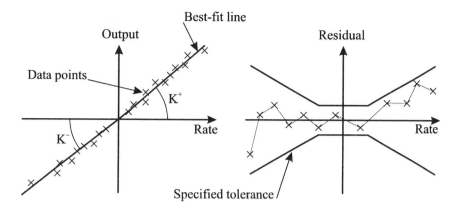

Figure 15.6. Scale factor measurement.

is to mount the gyro on a slip table, a horizontal plate on an oil bearing. The gyro is run closed-loop and the output signal is recorded with the vibration on and off. The vibration axis is placed in different orientations around the OA; midway between the IA and SA, and so on. The data reduction procedure for deriving D_{IS}, D_{SS}, and so on, is well described in IEEE Standard 517.

One might test strapdown SDFGs further [8] for response to angular acceleration about the OA, for anisoinertia, and the dependence of anisoinertia on SA oscillation frequency. These tests would usually be part of qualification testing rather than part of an ATP.

Testing the Dynamically Tuned Gyro

DTGs are tested very much as SDFGs are, except that they have two axes rather than one. To derive the drift coefficients we will again base our description on the IEEE approach [9] (Chapter 9) and write a simplified static error model for the x- and y-axes in terms of the scale factors K_x, K_y and the output signals S_x, S_y:

$S_x/K_x = \Omega_x$	Inertial rate
$+ B_x$	Bias, non-g-sensitive
$+ D_{xx}a_x$	Normal mass unbalance
$+ D_{xy}a_y$	Quadrature mass unbalance

$$S_y/K_y = \Omega_y \qquad\qquad \text{Inertial rate}$$

$$+ B_y \qquad\qquad \text{Bias, non-}g\text{-sensitive}$$

$$+ D_{yy}a_y \qquad\qquad \text{Normal mass unbalance}$$

$$+ D_{yx}a_x \qquad\qquad \text{Quadrature mass unbalance}$$

The D_{ab} coefficients can be read as "the drift on axis a caused by unit acceleration along axis b." Accelerations a_x, a_y include gravity.

We can measure these model coefficients using the eight-position test described next; repeating this test periodically will tell us their long-term stability, and we can enclose the test stand in a thermal chamber to find their temperature sensitivities.

The Eight-Position Test.
In this static test the gyro is moved through eight positions, and readings of x and y output are taken at each position. The eight positions are illustrated in Figure 15.7; they are in two groups of four. In the first group the gyro is positioned with the SA up, and data are taken with the x-axis pointing North, East, South, and West. Then the gyro is turned so that its SA points South, and the x-axis is pointed down, East, up, and West. We choose this sequence so that the gyro is not subjected to rotations about the IA (other than earth's rate) during the test, except when moving from position 4 to position 5, which must be done carefully so as not to exceed the servos' input rate range.

One selects the data-taking time at each position so as to obtain sufficient accuracy in the coefficients being determined, which will depend on the gyro random noise and the loop bandwidth. Typically a couple of minutes at each position will give sufficient accuracy for tactical gyros.

This test separates the non-g-sensitive and g-sensitive terms, as the first four positions do not point either x- or y-axis up or down. Thus we get bias and low rate scale factor (using earth's rate as input) from the first four positions, and from the second four we get another reading of bias and the mass unbalances. The data can also supply the axes' misalignments. We must correct our local value of gravitational acceleration to the standard value if we are measuring mass unbalance coefficients to sufficient precision.

Assume for simplicity that the gyro axes are orthogonal and aligned to the test table, which is itself level and aligned to True North. Writing the x and y output signals at position n (x_n, y_n) as a function of the input rates Ω_{eh} and Ω_{ev} (page 252) we get:

Position 1: $K_x x_1 = B_x + \Omega_{eh}$; $\qquad\qquad K_y y_1 = B_y$;

Position 2: $K_x x_2 = B_x$; $\qquad\qquad K_y y_2 = B_y - \Omega_{eh}$;

Position 3: $K_x x_3 = B_x - \Omega_{eh}$; $\qquad\qquad K_y y_3 = B_y$;

Position 4: $K_x x_4 = B_x$; $\qquad\qquad K_y y_4 = B_y + \Omega_{eh}$;

Position 5: $K_x x_5 = B_x + D_{xx} - \Omega_{ev}$; $\qquad\qquad K_y y_5 = B_y + D_{yx}$;

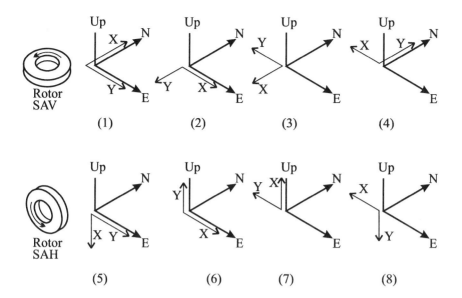

Figure 15.7. Eight-position test sequence.

Position 6: $K_x x_6 = B_x - D_{xy}$; $K_y y_6 = B_y - D_{yy} + \Omega_{ev}$;

Position 7: $K_x x_7 = B_x - D_{xx} + \Omega_{ev}$; $K_y y_7 = B_y - D_{yx}$;

Position 8: $K_x x_8 = B_x + D_{xy}$; $K_y y_8 = B_y + D_{yy} - \Omega_{ev}$;

from which

$$K_x = 2\Omega_{eh}/(x_1 - x_3);$$

$$K_y = 2\Omega_{eh}/(y_4 - y_2);$$

$$B_x = \tfrac{1}{2}K_x(x_1 + x_3) = \tfrac{1}{2}K_x(x_2 + x_4) = \tfrac{1}{2}K_x(x_5 + x_7)$$
$$= \tfrac{1}{2}K_x(x_6 + x_8);$$

$$B_y = \tfrac{1}{2}K_y(y_1 + y_3) = \tfrac{1}{2}K_y(y_2 + y_4) = \tfrac{1}{2}K_y(y_5 + y_7)$$
$$= \tfrac{1}{2}K_y(y_6 + y_8);$$

$$D_{xx} = \Omega_{ev} - \tfrac{1}{2}K_x(x_7 - x_5);$$

$$D_{yy} = \Omega_{ev} - \tfrac{1}{2}K_y(y_6 - y_8);$$

$$D_{xy} = \tfrac{1}{2}K_x(x_8 - x_6);$$

$$D_{yx} = \tfrac{1}{2}K_y(y_5 \quad y_7);$$

The SA-up biases should be identical to the SA-horizontal values, to within the measurement accuracy and the in-run random drift, but they might not be if the gyro has a sensitivity to acceleration along the SA. These sensitivities could be added to the model equation as coefficients D_{xz}, D_{yz} (Chapter 9).

DTG Rate Testing.
When rate testing a DTG, follow the procedure for the SDFG. Align one of the IAs with the servo table axis, keeping the SA horizontal and adjusting the alignment so that the second IA (perpendicular to the table axis) sees no table rate. The second IA operates in a closed loop.

Because DTGs have difficulty dissipating heat from their torquer windings, testing is more arduous for DTG torquers than actual use. While vehicles may have transient high rates, as in an airplane banking to turn, very few vehicles continuously turn at a high rate. Under test, though, the gyro must be rotated long enough to make an accurate measurement. Consequently the test must be planned so that it derives the scale factors, nonlinearities, asymmetries, and temperature dependencies of the X and Y torquers without overheating the instrument.

Testing Optical Gyros

Optical gyros can be tested like mechanical gyros with a few exceptions [10]. One can test their sensitivity to input voltage variation, to temperature, and to magnetic field; RLGs and FOGs can be sensitive to magnetic fields by the Zeeman and Faraday effects, respectively. They will be vibration and shock tested, because even though their drift is insensitive to acceleration, they can break.

But one must be cautious when testing the dithered RLG (Chapter 13), where the dither motion can interact with environmental vibration and give drift errors. For this reason, the dithered RLG must be tested on a rigid base, and only one gyro should be on the test stand at a time or they may interact.

Gyro testers can view optical gyros as rate gyros; they do not have torquers available to the user for driving a servo table. Therefore scale factor tests are carried out by measuring gyro output at fixed speeds, although the data are reduced in the same way as for mechanical gyros.

The Sigma Plot.
We saw in Chapter 11 that the rate random noise from optical gyros has a white spectrum. Thus, over a range of sample times, the RLG and the FOG will have a standard deviation (σ) that falls as the square root of the sample time (t). When integrated, the rate noise appears as random walk in angle (or angle random walk, ARW), a parameter independent of sample time.

A log-log plot of σ versus t is called a *Green chart* [11] or a *sigma plot* [12]. Figure 15.8 shows the sigma plot of a hypothetical gyro; it has three segments. At low sample times, the angle quantization of the output data dominates, falling as $1/t$. The random walk contribution falls as $1/\sqrt{t}$ so that it has a slope of -½ on the log-log plot. The right-most segment, due to rate random walk (deg/h/$\sqrt{h}$), comes

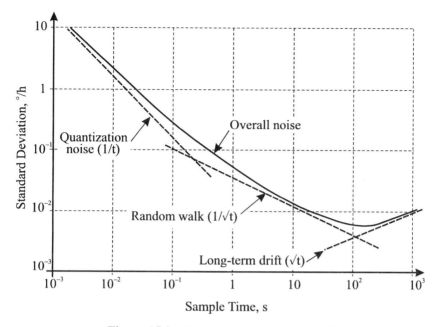

Figure 15.8. The sigma plot (Green chart).

from the long term drift processes, which begin to dominate at longer sample times, due, for example, to temperature variations. Matthews et al. use this approach to describe the Raytheon RLG [13].

The IEEE RLG test model [14] pays attention to absolute temperature and temperature gradient:

$$[P_o + P(\Omega) + P(T - T_o) + P(\Delta T)] \, N/t =$$

B	Bias
+ D(T − T$_o$)	Absolute temperature sensitivity
+ D(ΔT)	Gradient sensitivity
+ D$_R$	Random drift
+ Ω	Inertial rate about the IA

where
  P_o = pulse weight, arc-sec/pulse = 1/K
  K = scale factor
  $P(\Omega)$ = nonlinearity
  $P(T - T_o)$ = pulse weight versus absolute temperature
  $P(\Delta T)$ = pulse weight versus temperature gradient
  N = number of pulses
  t = sample time

The random walk in angle, R_θ is related to the r.m.s. value of D_R:

$$D_R = R_\theta/\sqrt{t}$$

Quantization error is omitted in this model, but must be considered in RLG testing, because, as we saw in Chapter 13, it can amount to an uncertainty of 1 deg/h for $t = 1$ s even for a navigation-grade gyro, forcing us to take 100-s samples in test, pushing test times into hours. Fortunately we can improve matters by hardware and software procedures; hardware can interpolate between pulses (for a fixed input rate) with a high-frequency clock and get arbitrarily small quantization. (This is not useful in a system, though, since it assumes that the rate is constant between pulses.) We can use software data processing procedures such as the moving average prefilter [15], which can differentiate between quantization noise and random walk.

The sigma plot can be useful when a system must operate in different modes; in the alignment mode it will be gyrocompassing, when longer samples can be taken. After that, the system might provide autopilot signals, which need high bandwidth and therefore short samples. The sigma plot provides this data while a single number may not, be it standard deviation or random walk.

Also, the industry will again benefit from the work of the Gyro and Accelerometer Panel of the IEEE, which has recently published a test procedure for Interferometric Fiber-Optic Gyros [16].

Conclusion

The descriptions in this chapter give some basic ideas of what is involved in testing accelerometers and gyros, both in equipment and procedure. If one has to set up test procedures in earnest, one should first consult the IEEE standards and then discuss the test plans with the customer, because it is quite likely that the customer has definite ideas on what is wanted.

References

1.   Lorenzini, D.A., "A new gyroscope centrifuge test," DGON Symposium Gyro Technology, 1972.
2.   Ingold, N.L., "Reverse velocity rocket sled test bed for inertial guidance systems," Navigation, J. Inst. Nav., 30, 1, pp. 90–99, Spring 1983.
3.   Peters, R.B., "Use of multiple acceleration references to obtain high precision centrifuge data at low cost," *Mechanical Technology of Inertial Devices*, Proc. Inst. Mech. Eng. (London), 2, Paper C47/87, 1987.
4.   IEEE-STD 836-1991. Precision Centrifuge Testing of Linear Accelerometers.

5.   IEEE STD 293-1969. Test Procedure for Single-Degree-of-Freedom Spring Restrained Rate Gyros.
6.   IEEE STD 517-1974. Standard Specification Format Guide and Test Procedure for Single Degree of Freedom Rate-Integrating Gyros.
7.   Shuttlewood, G.D., "Testing of inertial quality gyroscopes," *Symposium on Gyros*, Proc. Inst. Mech. Eng. (London), Vol. 179, 3E, 1964–5.
8.   IEEE STD 529-1980. Supplement for Strapdown Applications to IEEE Standard Specification Format Guide and Test Procedure for Single Degree of Freedom Rate-Integrating Gyros.
9.   IEEE STD 813-1988. Specification Format Guide and Test Procedure for Two-Degree-of-Freedom Dynamically Tuned Gyros.
10. Dyrna, P., "Test methods and test facilities to determine fiber optical gyro characteristics at IABG Inertial Test Laboratory," DGON Symposium Gyro Technology, Stuttgart, 1988.
11. Kochakian, C.R., "Time domain uncertainty charts (Green charts): A tool for validating the design of IMU/instrument interfaces," Proc. AIAA Guidance and Control Conference, Danvers, Mass., 1980.
12. King, A.D., "Characterisation of gyro in-run drift," DGON Symposium Gyro Technology, Stuttgart, 1984.
13. Matthews, J.B., M.I. Gneses, D.S. Berg, "A high-resolution laser gyro," Proc. IEEE NAECON, pp. 556–568, May 1978.
14. IEEE STD 647-1981. Specification Format Guide and Test Procedure for Single-Axis Laser Gyros.
15. Mark, J., A. Brown, "Laser gyroscope random walk determination using a fast filtering technique," DGON Symposium Gyro Technology, Stuttgart, 1984.
16. IEEE STD 952-1997. Specification Format Guide and Test Procedure for Single-Axis Interferometric Fiber-Optic Gyros.

16
Design Choices for Inertial Instruments

In this chapter we will summarize the tradeoffs between inertial systems designs and sensors, considering performance, cost, and reliability. The technology appropriate for a particular system depends on the performance and reliability needed and the size available; the cost will always need to be the lowest possible. We will also try to predict where the instrument design field is headed, allowing for the full impact of the Global Positioning System (GPS) satellite radio aid.

A Platform or a Strapdown System?

The performance needed will dictate the choice between the platform inertial measuring unit (IMU) and the strapdown type, and the nature of the mission will determine if radio or star-tracker aiding is permissible.

Platforms can have lower errors than strapdown systems. The null operation of their gyros eliminates anisoinertia and angular acceleration errors; spinning wheel gyros do not need to have high-power torquers whose dissipation causes thermal gradients. Missions demanding high accuracy and providing a benign environment might allow a gravity gradiometer to be used in a platform to compensate for gravity anomalies (Chapter 1); nuclear submarines typify this class. Platform accelerations can be measured with a Pendulous Integrating Gyro Accelerometer (PIGA, Chapter 7), currently the most precise accelerometer available, but the PIGA cannot be used in a strapdown system.

To summarize the other benefits of gimballed platforms:

1. they can operate with vehicle rotation rates greater than 1000 deg/s,
2. they self-align by gyrocompassing,
3. they can calibrate the sensors by platform rotations, and
4. gyro torquer errors do not lead to attitude error.

But they are mechanically complicated, with their slip rings and bearings, synchros, and torque motors. Accordingly, platforms are larger, more expensive, and have lower reliability than strapdown systems. The Advanced Inertial Reference System (AIRS), used in the MX strategic missile, replaces the gimbal set with flotation; the gyros and accelerometers are assembled into a ball that is neutrally buoyant in a fluid (like the gimbal in floated gyros, Chapters 7 and 8). However, the complexity of the usual gimbals and bearings is equalled by that of the fluidic torque motors, fluid pumps, heat exchangers, and so on needed for its operation.

Conversely, strapdown systems are lighter, simpler, cheaper, more easily configured for odd-shaped spaces, and due to the absence of moving parts, they can have high reliability. Their deficiencies are:

1. with spinning-wheel gyros (but not optical gyros) they have practical limits to maximum rate, in the 400–600 deg/s range,
2. they are difficult to align,
3. the sensors cannot easily be calibrated so must be stable,
4. with spinning-wheel gyros (but not optical gyros) and with pendulous accelerometers, system rotation induces sensor errors (torquer asymmetry, nonlinearity, anisoinertia, and angular acceleration), and
5. their accelerometer bias errors accumulate.

Optical gyros (RLG, IFOG) are naturally the best choice for strapdown systems.

Aiding the IMU

If the IMU can incorporate a star-tracker or a radio aid, the integrated system will have better performance at lower cost. Strategic weapons systems must remain self-contained, perhaps aided with a star-tracker, because in a war serious enough for strategic weapons to be used one could not be sure that radio aids, like the constellation of GPS satellites, would survive. They could either be destroyed, or their signals could be jammed.

For commercial and tactical military use, GPS aiding has changed the nature of the navigation field. The lower cost of GPS/inertial systems is expanding automatic navigation. Aided systems are generally strapdown systems, because GPS and strapdown systems provide complementary performance, and the two can be integrated in a number of ways [1]. In one scheme, a common processor can service both the strapdown and the GPS, and the IN system can tell the GPS receiver where to look for satellites and maintain navigation while satellites are jammed or obscured by buildings or passage through a tunnel.

In any case, the GPS receiver updates the IN system regularly, so that the INS can use instruments with lower day-to-day and in-run drift and relaxed scale factor. These lower performance gyros and accelerometers are expected to be considerably less expensive, by as much as an order of magnitude.

GPS helps in surveying test sites, and differential systems using both GPS and GLONASS satellites can achieve centimeter surveying accuracy on test ranges.

Choice of Sensor Type

In earlier chapters we described sensors by case studies of manufactured models. It is instructive to categorize them in other ways; one can distinguish, for example, between sensors that use differential design and those that use resonance. Then again, one can choose between optical and mechanical gyros.

Differential Design

In the open-loop accelerometer or the single-axis rate gyro, the displacement of the sensitive element signals the input, and consequently the sensor suffers from cross-coupling and vibropendulous errors. We avoid this in the closed-loop instrument by operating at null; the pickoff signal (proportional to the difference between the acceleration or rate input torque and the feedback torque) is zero in the steady state.

The vibrating beam accelerometer (Chapter 5), the Mach-Zehnder, Sagnac, and Fabry-Perot interferometers (Chapter 11) all exhibit differential operation. Oscillators are connected in push-pull, or paths are common and change length by opposite amounts in response to the inertial stimulus. This symmetry or reciprocity provides high rejection of the common mode signal (the unloaded tuning fork frequency, the total path phase) so that the sensor has high resolution. For example, the CW and CCW frequencies of a ring laser gyro (Chapter 13) are of the order of 10^{14} Hz, yet their difference can be read to 10^{-3} Hz.

Using Resonance

Uncontrolled resonance of mechanical parts can be destructive to systems and sensors. The dynamically tuned gyro's rotor (Chapter 9) is suspended on elastic flexures on a shaft, in a very low damping environment. Under external vibration at the resonance of the rotor mass M_r on the net flexure bending stiffness K_b, $\omega = (K_b/M_r)^{1/2}$, the stress in the flexures can exceed the elastic limit. Drift errors from g and g^2 terms are higher at resonance. Clearly, the natural frequency will be designed to be higher than the environmental frequencies; antivibration mounts may even be used to support the gyro.

But resonance is also used constructively in the DTG and other sensors. In the DTG, the gyroscopic torque from gimbal angular oscillation (the dynamic anti-spring) cancels the positive torque of the flexure twist at the tuned speed. The tuning forks of the VBA and the Quartz Rate Sensor (Chapter 10), and the resonator of the Hemispherical Resonator Gyro (Chapter 10) operate at a resonance so that the amplitude of motion is large; the high Q-factor provides high resolution of the output frequency.

The RLG and the other resonant optical sensors, Resonant Fiber Optic Gyro and Micro Optic Gyro (Chapters 13 and 14) accumulate Sagnac phase because resonance causes multiple circuits of the resonator perimeter; the finesse (like the

Q-factor) expresses the sharpness of resonance and the resolution. The nonresonant Interferometric Fiber-Optic Gyro (Chapter 12) achieves the same resolution by using a long, coiled, fiber path rather than resonance.

Mechanical or Optical Gyros?

The differences between optical and mechanical gyros affect their selection for particular applications. Mechanical gyros provide lower short-term noise than optical gyros (but worse long-term stability), they have inertial memory, and they can offer longer life.

Performance: Mechanical gyros with wheels, and the vibrating HRG, have low short-term noise, and gas bearing single-degree-of-freedom gyros (Chapter 7) have demonstrated output noise as low as a few nanoradians. They are preferable for the precise pointing of telescopes and laser designators for short times.

On the other hand, optical gyros have no g-sensitivity, can measure high rates without an increase in power, have better day-to-day drift stability, and have stable, linear scale factor.

Inertial Memory.
Both wheeled and vibrating gyros have inertial memory. If power is lost for a short time (up to a few seconds), the sensitive element integrates the input motions so that when power returns, the gyro signals the angle turned in the interval. For gimballed wheels (either SDFG or DTG) this angle must not be so large that the gimbal or rotor hits a limit stop, practically 1–3°. Nor, for the HRG, can the rotation be so large that one or more nodes have passed the pickoff (nodes being indistinguishable from one another). The ESG (Chapter 8), on the other hand, will catastrophically crash if it loses suspension power in an accelerating environment (including gravity), but otherwise it will have very long inertial memory.

Optical gyros do not have this memory feature; once the power is off, the gyro loses reference. Optical gyros can also lose memory under nuclear radiation, either from super-radiance in the laser (short term) or from blackening of the sensing and Er-doped source fibers (long term).

Lifetime.
Gas bearing SDFGs have very long lifetimes, particularly if the wheel is kept running, because there is no material contact in the bearing and no wearing. The same is true of the ESG. The RLG depends on the correct amount of helium for its operation, and helium is noted for its ability to diffuse through many materials, which can limit RLG life. The IFOG, a solid-state gyro, has no such limitation, and should be very reliable. For the vibrating sensors, the QRS lifetime should be unlimited, although the performance of the HRG depends on maintaining a hard vacuum around the resonator. Gettering can remove outgassed vapors in order to extend life.

Reliability

Reliability is the probability that a system will perform for a given time under given conditions. The system is assumed to have been randomly chosen from a group of identical systems; reliability is a characteristic of the group, not of the individual. The reliability of inertial sensors is often estimated by similarity—we have numbers on this old model and the new one looks similar therefore its reliability must be the same—replaced by test data as early as possible.

Experience has shown that some systems tend to fail early, so to weed out early failures it is common to burn-in systems for a week or so.

Things often have a finite life. Some gracefully degrade, others catastrophically fail. Bearings and slip rings wear, gas laser gyro cathodes may sputter away, and a DTG flexure or an accelerometer hinge might break. These failures might call for maintenance (the process of returning the system as closely as possible to its as-new condition), if the system was designed to be repairable. When it is not economically feasible to repair a system, it has worn out. The mean life of a population of systems is the average time to wearout, and the useful life of a system is the period between burn-in and the onset of wearout.

If we can repair a failed device so that the repair does not introduce a new failure potential, then the average time that one can expect the device to work before the next failure is the *mean time between failures*. In a test of a large sample, the MTBF is the total operating time of all the samples (not just the failed ones) divided by the number of failures. After a long time, this test should converge on a constant MTBF. The ratio of MTBF to mission time is the probability that the device will allow the mission to be completed. Statistically, 63% of a population of a device will fail in an operating time equal to the MTBF. Failure rate is the reciprocal of the MTBF and is usually expressed in failures per 10^6 hours.

Redundancy

In order to ensure a successful mission, any single system that could cause the mission to fail must have an adequate MTBF. If it does not, or if the consequence of any failure is unacceptable, backup systems will provide redundancy. Redundancy can be provided by installing a few independent, identical copies of a system and comparing their outputs. Inertial navigation systems are often used in threes, and a computer correlates their outputs (as well as monitoring obvious failures like a blown fuse). If two of the three consistently differ from the third, the third is considered to have failed and its data are ignored. Deciding between two might be possible if other data from separate instruments are available.

Redundant systems are expensive, and the cost of providing backup systems must include the cost of carrying their extra weight. Rather than installing three copies of a system, redundancy can be provided at the component level. Particular system designs include one in which a fourth single-axis sensor is mounted with its sensitive axis skewed to the three basic sensor axes. This sensor can then act as a check on the basic three and can provide information if one is determined to have failed.

Sensor Design Check Lists

Much of the information in this book can be summarized into check lists, and Tables 16.1 and 16.2 list many of the performance and environmental factors to be considered in applying inertial sensors. Table 16.3 shows how different performance features vary with gyro type.

Table 16.1.   Sensor selection factors—performance.

Operating life, MTBF	OA acceleration sensitivity
Activation time	1N and 2N sensitivities
Maximum rate or acceleration	Transfer characteristics:
Hysteresis	bandwidth
Scale factor: stability, long term	damping
linearity and asymmetry	Thermal: temperature sensitivities
Anisoinertia	maximum operating temperature
Axis alignment stability	minimum operating temperature
Cross-coupling	gradients
Threshold and resolution	shocks
Bias, day-to-day repeatability	Magnetic field sensitivity
in-run random	Resonant frequencies
random walk	Power consumption
Noise spectrum	unusual voltages/frequencies
g-sensitivity	Size and mass
Anisoelasticity (g^2 terms)	Self-test
Vibropendulosity	

Table 16.2.   Sensor selection factors—environmental.

Storage life	Shock: wave shape
Temperature extremes	peak value
Linear acceleration, level	duration
duration	(e.g., 50g half-sine, 15 ms)
directions	Shock mount: transmissibility
Vibration spectra: sine sweep	resonances
frequency range	axis alignment degradation
r.m.s. level	Air pressure/vacuum
random, g^2/Hz profile	Ionizing radiation
Nonoperating resonances (fatigue)	

Table 16.3. Performance dependencies of some gyros.

Feature	SDFG	DTG	FOG/MOG	RLG
Maximum rate	Torquer heat dissipation		Not instrument limited, to first order	
Scale factor stability	Permanent magnet aging		Wavelength stability	
Scale factor linearity	Quality of magnetic design		Quality of feedback system (serrodyne)	Lock-in rate and means of suppression
Bias stability (long-term)	Flexleads Magnetic leakages	Time const. Mistuning Magnetic leakage Bearing noise		Backscatter and mode control
Random drift (short-term bias)	Fluid OA bearings Spin bearings	Spin bearings Gas drag pickoff stability	Quantum shot noise	Spontaneous emission Mechanical dither Quantization
g-sensitivity	Gimbal mass stability	Rotor mass stability Flexure quadrature torque	None to the first order	
IA stability	OA bearing Pickoff null Case-mounting flange	Pickoff nulls Case-mounting flange	FOG coil former MOG chip mounting	Dither motor

Conclusions

There are few choices for inertial sensors for the ultimate performance, with better than 1 mile/day system drift. The field is limited to platforms carrying the SDFG gas bearing model, the ESG, the continuously rotated RLG, and the PIGA. Gyro technology does not currently offer new technology for this market although the HRG and the IFOG are making progress. And, there is no accelerometer yet that is capable of replacing the PIGA—the world definitely needs a smaller, less expensive, more reliable, accelerometer of PIGA performance.

But for low-accuracy systems, sensor choices are not clear-cut. The least expensive inertial sensors are typified by the quartz VBA and the QRS tuning fork gyro, and the silicon-based micromachined silicon accelerometer and tuning fork gyros, all too new to have pedigree. Their selection implies some risk.

For new medium-accuracy GPS/IN systems, Heading and Attitude Reference Systems, and the like, the IFOG and the VBA should be chosen. Although the SDFG and DTG have the pedigree and an extensive installed base, they are less attractive because of their high rate power dissipation, lower reliability (from the ball bearings), and higher cost.

Reference

1. Johannessen, R., M.J.A. Asbury, "Towards a quantitative assessment of benefits INS/GPS integration can offer to civil aviation," Navigation, 37, 4, pp. 329–346, Winter 1990–91.

Index

Laminar Viscous Flow
V.N. Constantinescu

Thermal Contact Conductance
C.V. Madhusudana

Transport Phenomena with Drops and Bubbles
S.S. Sadhal, P.S. Ayyaswamy, and J.N. Chung

**Fundamentals of Robotic Mechanical Systems:
Theory, Methods, and Algorithms**
J. Angeles

Electromagnetics and Calculations of Fields
J. Ida and J.P.A. Bastos

Mechanics and Control of Robots
K.C. Gupta

**Wave Propagation in Structures:
Spectral Analysis Using Fast Discrete Fourier Transforms, 2nd ed.**
J.F. Doyle

Fracture Mechanics
D.P. Miannay

Principles of Analytical System Dynamics
R.A. Layton

**Composite Materials:
Mechanical Behavior and Structural Analysis**
J.M. Berthelot

**Modern Inertial Technology:
Navigation, Guidance, and Control, 2nd ed.**
A. Lawrence

**Dynamics and Control of Structures:
A Modal Approach**
W.K. Gawronski